Lecture Notes in Bioinformatics

Subseries of Lecture Notes in Computer Science

Corrado Priami Falko Dressler
Ozgur B. Akan Alioune Ngom (Eds.)

Transactions on Computational Systems Biology X

Springer

Series Editors

Sorin Istrail, Brown University, Providence, RI, USA
Pavel Pevzner, University of California, San Diego, CA, USA
Michael Waterman, University of Southern California, Los Angeles, CA, USA

Editor-in-Chief

Corrado Priami
The Microsoft Research - University of Trento
Centre for Computational and Systems Biology
Piazza Manci, 17, 38050 Povo (TN), Italy
E-mail: priami@cosbi.eu

Guest Editors

Falko Dressler
University of Erlangen
Department of Computer Science 7
Martensstr.3, 91058 Erlangen, Germany
E-mail: dressler@informatik.uni-erlangen.de

Ozgur B. Akan
Middle East Technical University
Department of Electrical and Electronics Engineering
Ankara, Turkey 06531
E-mail: akan@eee.metu.edu.tr

Alioune Ngom
University of Windsor
School of Computer Science
5115 Lambton Tower
401 Sunset Avenue, N9B 3P4 Windsor, ON, Canada
E-mail: angom@uwindsor.ca

Library of Congress Control Number: Applied for

CR Subject Classification (1998): J.3, F.1-2, F.4, I.6, C.2

LNCS Sublibrary: SL 8 – Bioinformatics

ISSN	0302-9743 (Lecture Notes in Computer Science)
ISSN	1861-2075 (Transactions on Computational Systems Biology)
ISBN-10	3-540-92272-5 Springer Berlin Heidelberg New York
ISBN-13	978-3-540-92272-8 Springer Berlin Heidelberg New York

springer.com

© Springer-Verlag Berlin Heidelberg 2008
Printed in Germany

Typesetting: Camera-ready by author, data conversion by Scientific Publishing Services, Chennai, India
Printed on acid-free paper   SPIN: 12589234   06/3180   5 4 3 2 1 0

Preface

Technology is taking us to a world where myriads of heavily networked devices interact with the physical world in multiple ways, and at many levels, from the global Internet down to micro and nano devices. Many of these devices are highly mobile and autonomous and must adapt to the surrounding environment in a totally unsupervised way.

A fundamental research challenge is the design of robust decentralized computing systems that are capable of operating in changing environments and with noisy input, and yet exhibit the desired behavior and response time, under constraints such as energy consumption, size, and processing power. These systems should be able to adapt and learn how to react to unforeseen scenarios as well as to display properties comparable to social entities. The observation of nature has brought us many great and unforeseen concepts. Biological systems are able to handle many of these challenges with an elegance and efficiency far beyond current human artifacts. Based on this observation, bio-inspired approaches have been proposed as a means of handling the complexity of such systems. The goal is to obtain methods to engineer technical systems, which are of a stability and efficiency comparable to those found in biological entities.

This *Special Issue on Biological and Biologically-inspired Communication* contains the best papers from the Second International Conference on Bio-Inspired Models of Network, Information, and Computing Systems (BIONETICS 2007). The BIONETICS conference aims to bring together researchers and scientists from several disciplines in computer science and engineering where bio-inspired methods are investigated, as well as from bioinformatics, to deepen the information exchange and collaboration among the different communities.

We selected eight outstanding papers from both domains and invited the authors to prepare extended versions of their manuscripts. The first three papers by Forestiero et al., Meyer et al., and Dubrova et al. describe the applicability of bio-inspired techniques in the technical domain of computing and communication. Atakan and Akan, and Nakano et al. focus on molecular communication and the properties of such communication channels. In the domain of bioinformatics, the papers by Rueda et al., and Smith and Wiese demonstrate techniques for the analysis of genes. We would like to highlight the paper by Wang et al. in which a bio-inspired approach is applied in the field of bioinformatics. This approach specifically complements our special issue on bio-inspired solutions and bioinformatics. The final paper describes a stochastic pi-calculus model of the PHO pathway, and it is a regular paper.

July 2008

Falko Dressler
Özgür B. Akan
Alioune Ngom

LNCS Transactions on Computational Systems Biology – Editorial Board

Table of Contents

Biological and Biologically-Inspired Communication

Towards a Self-structured Grid: An Ant-Inspired P2P Algorithm*

Agostino Forestiero, Carlo Mastroianni,
Giuseppe Papuzzo, and Giandomenico Spezzano

ICAR-CNR
Via P. Bucci, 41C, 87036 Rende(CS), Italy
{forestiero,mastroianni,papuzzo,spezzano}@icar.cnr.it

Abstract. This paper introduces Antares, a bio-inspired algorithm that exploits ant-like agents to build a P2P information system in Grids. The work of agents is tailored to the controlled replication and relocation of metadata documents that describe Grid resources. These descriptors are indexed through binary strings that can either represent topics of interest, specifically in the case that resources are text documents, or be the result of the application of a locality preserving hash function, that maps similar resources into similar keys. Agents travel the Grid through P2P interconnections and, by the application of ad hoc probability functions, they copy and move descriptors so as to locate descriptors represented by identical or similar keys into neighbor Grid hosts. The resulting information system is here referred to as *self-structured*, because it exploits the self-organizing characteristics of ant-inspired agents, and also because the association of descriptors to hosts is not pre-determined but easily adapts to the varying conditions of the Grid. This self-structured organization combines the benefits of both *unstructured* and *structured* P2P information systems. Indeed, being basically unstructured, Antares is easy to maintain in a dynamic Grid, in which joins and departs of hosts can be frequent events. On the other hand, the aggregation and spatial ordering of descriptors can improve the rapidity and effectiveness of discovery operations, and also enables range queries, which are beneficial features typical of structured systems.

Keywords: Ant Algorithms, Grid, Information Dissemination, Information System, Peer-to-Peer.

1   Introduction

Grid computing [14] is an emerging computing model that provides the ability to perform higher throughput computing by taking advantage of many networked computers and distributing process execution across a parallel infrastructure.

---

* This paper is an extended version of the paper A. Forestiero, C. Mastroianni, G. Spezzano, "*Antares: an Ant-Inspired P2P Information System for a Self-Structured Grid*". BIONETICS 2007 - 2nd International Conference on Bio-Inspired Models of Network, Information, and Computing Systems, Budapest, Hungary, December 2007.

C. Priami et al. (Eds.): Trans. on Comput. Syst. Biol. X, LNBI 5410, pp. 1–19, 2008.

The *information system* is an important pillar of a Grid framework, since it provides information that is critical to the operation of the Grid and the construction of applications. In particular, users turn to the information system to discover suitable resources or services that are needed to design and execute a distributed application, explore the properties of such resources and monitor their availability.

Due to the inherent scalability and robustness of P2P algorithms, several P2P approaches have been recently proposed for resource organization and discovery in Grid environments [15]. The ultimate goal of these approaches is to allow users to rapidly locate Grid resources or services (either hardware or software) which have the required characteristics; this is generally reduced to the problem of finding related *descriptors*, through which it is possible to access the corresponding resources. A descriptor may contain a syntactical description of the resource/service (i.e. a WSDL - Web Services Description Language - document) and/or an ontology description of resource/service capabilities. In a Grid, after issuing a query, a user can discover a number of descriptors of possibly useful resources, and then can choose the resources which are the most appropriate for their purposes.

In P2P systems, descriptors are often indexed through bit strings, or *keys*, that can have two different meanings. The first is that each bit represents the presence or absence of a given *topic* [8] [19]: this method is particularly used if the resource of interest is a document, because it is possible to define the different topics on which this document focuses. Alternatively, a resource or service (for example a computation resource) can be mapped by a hash function into a binary string. The hash function is assumed to be locality preserving [5] [17], which assures that resources having similar characteristics are associated to similar descriptor keys. Similarity between two resources can be measured as the cosine of the angle between the bit vectors through which the corresponding descriptors are indexed.

In this paper, we propose Antares (ANT-based Algorithm for RESource management in Grids) a novel approach for the construction of a Grid information system, which is inspired by the behavior of some species of ants [4]. The Antares algorithm is able to disseminate and reorganize descriptors and, as a consequence of this, it facilitates and speeds up discovery operations. More specifically, Antares concurrently achieves multiple objectives: (i) it *replicates* and *disseminates* descriptors on the network; (ii) it *spatially sorts* descriptors, so that descriptors indexed by similar keys are placed in neighbor hosts; (iii) thanks to the self-organizing nature of the ant-based approach, the reorganization of descriptors spontaneously adapts to the ever changing environment, for example to the joins and departs of Grid hosts and to the changing characteristics of resources.

The Grid information system constructed with Antares is basically *unstructured*, in the sense that descriptors are not required to be mapped onto specified hosts (for example, hosts determined by a hash function, as in most structured P2P systems), but they are freely placed by ants through their pick and drop operations. This assures valuable features such as self-organization, adaptivity, and ultimately scalability.

Nevertheless, Antares features a self-emerging organization of descriptors, which has some "structured" properties, since descriptors are aggregated and spatially sorted. Therefore, the resulting information system is referred to as *self-structured*, because it exploits the self-organizing characteristics of ant-inspired agents, and also because the association of descriptors to hosts is not predetermined but adapts to the varying conditions of the Grid.

In actual fact, Antares retains important benefits, which are typical of *structured* systems. In particular, it enables the use of an *informed* discovery algorithm, through which a query message, issued to discover resources indexed by a specified *target* descriptor, can efficiently explore the Grid and collect information about such resources. The discovery algorithm is based on a best neighbor approach: at each step, it drives the query message towards the neighbor peer that possesses descriptors which are the most similar to the target descriptor. Since descriptors have been spatially sorted by Antares agents, this algorithm allows the query to reach a Grid region in which several useful descriptors have been accumulated.

In this paper, we show that the Antares algorithm succeeds in the spatially replication and sorting of descriptors. In fact, event-based simulation proves that agents successfully generate and disseminate several replicas of each resource, and at the same time that the homogeneity of descriptors located in each small Grid region is notably increased, meaning that descriptors are effectively reorganized and sorted on the Grid.

2 Related Work

Management and discovery of resources in distributed systems such as Grids and P2P networks is becoming more and more troublesome due to the large variety and dynamic nature of hosts and resources. Centralized approaches are becoming unbearable, since they create administrative bottlenecks and hence are not scalable. Novel approaches for the construction of a scalable and efficient information system, need to have the following properties: self-organization (meaning that Grid components are autonomous and do not rely on any external supervisor), decentralization (decisions are to be taken only on the basis of local information) and adaptivity (mechanisms must be provided to cope with the dynamic characteristics of hosts and resources).

Requirements and properties of Self Organizing Grids are sketched in [10]. Some of the issues presented in this paper are concretely applied in our work: for example, reorganization of resources in order to facilitate discovery operations, and adaptive dissemination of information. Another self-organization mechanism is proposed in [1] to classify Grid nodes in groups on the basis of a similarity measure. Each group elects a leader node that receives requests tailored to the discovery of resources which are likely to be maintained by such group. This is an interesting approach but it still has non-scalable characteristics: for example, it is required that each Grid node has a link to all the leader nodes, which is clearly problematic in a very large Grid. A self-organizing mechanism is also exploited in [7] to build an adaptive overlay structure for the execution of a large number of tasks in a Grid.

Similarly to the latter work, the Antares algorithm presented in this paper exhibits several characteristics of both biological systems and mobile agent systems (MAS). Antares is specifically inspired to ant algorithms, a class of agent systems which aims to solve very complex problems by imitating the behavior of some species of ants [4]. Antares agents are simple and perform simple operations, but a sort of "swarm intelligence" emerges from their collective behavior. Moreover, agents are able to adapt their individual behavior autonomously, without having an overall view of the system, for example to select their mode of operation, between *copy* and *move*. This approach is similar to that discussed in [18], where a decentralized scheme, inspired by insect pheromone, is used to tune the activity of a single agent when it is no longer concurring to accomplish the system goal.

The Antares approach can be positioned along a well known research avenue whose objective is to devise possible applications of *ant algorithms*, i.e., algorithms inspired by the behavior of ants [4] [9]. An approach for the reorganization and discovery of resources that are pre-classified in a given number of classes has been recently proposed [12] [13]. This enables the creation of Grid regions specialized in a particular class, which improves the performance of discovery operations, but does not allow for the spatial sorting of descriptors, thus preventing the possibility of efficiently managing range queries.

Antares has been specifically designed to tackle the case in which the access keys of resource descriptors are bit strings and, since similar resources are assumed to be mapped into similar strings, it is possible to define a similarity measure among resources, through the comparison of related keys. In this sense our work is partly inspired by the work of Lumer and Faieta [16], who devised a method to spatially sort data items through the operations of simple robots. However, the approach of Lumer and Faieta has been adapted to our purposes, by making the following main modifications: (i) descriptors are not only sorted, as in [16], but also *replicated*, in order to disseminate useful information on the Grid and facilitate search requests; (ii) each peer can contain a number of descriptors, not just one item as in [16], thus enabling the accumulation of descriptors on Grid hosts; (iii) since Antares operates in a distributed computing environment, agents limit the number of P2P hops in order to reduce traffic and computing load.

Antares features two noteworthy properties, which are interesting enhancements with respect to most existing strategies: (i) it does not rely on any centralized support, but it is fully decentralized, self-organizing and scale-free, thanks to its bio-inspired nature and its swarm intelligence characteristics; (ii) being basically unstructured, it avoids the typical problems of P2P structured systems, but still retains some of their important benefits, such as the efficient management of queries, thus proposing itself as a good compromise between the two P2P strategies, structured and unstructured, which are generally deemed as completely alternative one to the other.

3   The Antares Algorithm

The main purpose of the Antares algorithm is to disseminate resource descriptors over the Grid and at the same time achieve a logical organization of Grid resources by spatially sorting the corresponding descriptors according to the their keys.

The Grid system uses P2P interconnections to enable communication and exchange of descriptors among Grid hosts. This is coherent with the recent trend of adopting P2P techniques in Grid frameworks, in order to enhance efficiency and scalability features of large-scale Grids [15] [21].

The Antares information system is progressively and continuously constructed by a number of ant-inspired agents which travel the Grid through these P2P interconnections, possibly *pick* resource descriptors from a Grid host, carry these descriptors, and *drop* them into other hosts. *Pick* and *drop* operations are based on the evaluation of the corresponding probability functions. Though these operations are very simple, and agents are unaware of the significance of what they do, a sort of *swarm intelligence* emerges from their combined work, which is typical of ant systems, and of bio-inspired systems in general.

A high-level description of the algorithm, performed by mobile agents, is given in the flowchart of Figure 1. Periodically, each agent performs a small number of P2P hops among Grid hosts. Whenever an agent arrives at a new Grid host, it evaluates the pick or drop probability function, specifically: (i) if the agent does not carry any descriptor, it evaluates the *pick probability function* for every resource descriptor stored in this host, so as to decide whether or not to pick these descriptors; (ii) if the agent already carries some descriptors, it evaluates the *drop probability function* for each of these descriptors, so as to decide whether or not to leave them in the current host. After picking some descriptors, the agent will carry them until it drops them into another host, and then will try to pick other descriptors from another host. An agent can replicate descriptors, or simply move them, depending on the "mode" with which it operates, the *copy* mode or *move* mode. This will be better discussed in the following.

3.1   Pick Operation

Periodically, an agent performs a small number of P2P hops among Grid hosts (see Figure 1). Whenever an agent arrives at a new Grid host, and it does not carry any descriptor, it evaluates the *pick probability function* and decides whether or not to pick one or more descriptors from the current host.

Specifically, the agent checks each single descriptor maintained in the current host, and evaluates its *average similarity* with all the descriptors maintained by the hosts located in the *visibility region*. The *visibility region* includes all the hosts that are located within the *visibility radius*, i.e., that are reachable

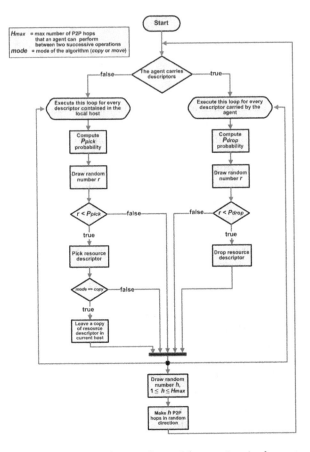

Fig. 1. The algorithm performed by ant-inspired agents

from the current host with a given number of hops. This radius is an algorithm parameter, and is set here to 1, in order to limit the amount of information exchanged among hosts.

Actually, the agent evaluates the similarity of the binary key of the descriptor under consideration with the keys of the *centroids* of the current host and of the neighbor hosts, and then takes the average. A *centroid* of a host is a virtual descriptor whose key is representative for the descriptors maintained in the local host. Similarity is evaluated against centroids (instead of against all single descriptors) in order to reduce the information exchanged among hosts. In fact each host must only know the keys of the centroids of the neighbor peers, instead of the keys of all the descriptors.

The probability of picking a descriptor must be inversely proportional to the average similarity of this descriptor with those located in the visibility region, thus obtaining the effect of averting a descriptor from co-located dissimilar descriptors. As soon as the possible initial equilibrium is broken (i.e., descriptors having different keys begin to be accumulated in different Grid regions), a further reorganization of descriptors is increasingly driven, because the probability of picking an "outlier" descriptor increases.

The pick probability function P_{pick} is defined in formula (1) whereas f, defined in formula (2), measures the average similarity of a generic descriptor d with the other descriptors located in the visibility region R; the value of f assumes values ranging between 0 and 1, and so does P_{pick}. [1].

In more detail, in formula (1) the parameter k_p, whose value is comprised between 0 and 1, can be tuned to modulate the degree of similarity. In fact, the pick probability is equal to 0.25 when f and k_p are comparable, while it approaches 1 when f is much lower than k_p (i.e., when the descriptor in question is extremely dissimilar from the others) and 0 when f is much larger than k_p (i.e., when the descriptor in question is very similar to the others). Here k_p is set to 0.1, as in [4].

$$Ppick = \left(\frac{k_p}{k_p + f}\right)^2 \tag{1}$$

In formula (2), the weight of each term is equal to the number of descriptors N_p maintained in each peer p, while N is the overall number of descriptors maintained in the region R, i.e., $N = \sum_{(p \in R)} N_p$. Note that the similarity between d and the descriptor of the centroid of the peer p, C_p, is defined as the cosine of the angle between the corresponding key vectors. The parameter α defines the similarity scale [16]; here it is set to 0.5.

$$f(d, R) = \frac{1}{N} \cdot \sum_{p \in R} N_p \cdot (1 - \frac{1 - cos(d, C_p)}{\alpha}) \tag{2}$$

After evaluating the pick probability function, the agent computes a random real number comprised between 0 and 1, then it executes the pick operation if this number is lower than the value of the pick function. As the local region accumulates descriptors having similar keys, it becomes more and more likely that "outlier" descriptors will be picked by an agent.

The *pick* operation can be performed with two different modes, *copy* and *move*. If the *copy* mode is used, the agent, when executing a pick operation, leaves the descriptor on the current host, *generates a replica* of it, and carries the new descriptor until it drops it into another host. Conversely, with the *move* mode, an agent picks the descriptor and *removes* it from the current host, thus preventing an excessive proliferation of replicas. These two modes and their impact are better discussed in Section 3.3.

3.2 Drop Operation

As well as the pick function, the *drop probability function* P_{drop} is first used to break the initial equilibrium and then to strengthen the spatial sorting of

---

[1] Actually, with the adopted values of α and k_p, the value of f can range between -1 and 1, but negative values are truncated to 0: this corresponds to have, in formula (1), a P_{pick} value of 1 in the case that the evaluated descriptor is very dissimilar from the other descriptors.

descriptors. Whenever an agent gets to a new Grid host, it must decide, if it is carrying some descriptors, whether or not to drop these descriptors in the current host.

For each carried descriptor, the agent separately evaluates the drop probability function, which, as opposed to the pick probability, is directly proportional to the similarity function f defined in formula (2), i.e., to the average similarity of this descriptor with the descriptors maintained in the current visibility region.

In (3), the parameter k_d is set to a higher value than k_p, specifically to 0.5, in order to limit the frequency of drop operations. Indeed, it was observed that if the drop probability function tends to be too high, it is difficult for an agent to carry a descriptor for an amount of time sufficient to move it into an appropriate Grid region.

$$Pdrop = \left(\frac{f}{k_d + f} \right)^2 \tag{3}$$

As for the pick operation, the agent first evaluates P_{drop}, then extracts a random real number between 0 and 1, and if the latter number is lower than P_{drop}, the agent drops the descriptor in question into the current host.

As a final remark concerning pick and drop probability functions, it is worth specifying that the values of mentioned parameters (k_p, k_d, α) have an impact on the velocity and duration of the transient phase of the Antares process, but they have little influence on the performance observed under steady conditions.

3.3   Spatial Sorting of Descriptors

The effectiveness of Antares is evaluated through a spatial *homogeneity function* H. Specifically, for each host of the Grid, we calculate the homogeneity among all the descriptors maintained within the local visibility region, by averaging the cosine of the angle between every couple of descriptors. Afterwards, the values of the homogeneity functions calculated on all the hosts of the network are averaged. The objective is to increase the homogeneity function as much as possible, because it would mean that similar descriptors are actually mapped and aggregated into neighbor hosts, and therefore an effective sorting of descriptors is achieved.

Simulation analysis showed that the overall homogeneity function is better increased if each agent works under both its operational modes, i.e., *copy* and *move*. In the first phase of its life, an agent is required to *copy* the descriptors that it picks from a Grid host, but when it realizes from its own activeness that the sorting process is at an advanced stage, it begins simply to *move* descriptors from one host to another, without creating new replicas. In fact, the *copy* mode cannot be maintained for a long time, since eventually every host would maintain a very large number of descriptors of all types, thus weakening the efficacy of spatial reorganization. The algorithm is effective only if each agent, after replicating a number of descriptors, switches from *copy* to *move*.

A self-organization approach based on the concept of *stigmergy* [6] enables each agent to perform this mode switch only on the basis of local information.

This approach is inspired by the observation that agents perform more operations when the system entropy is high (because descriptors are distributed randomly), but operation frequency gradually decreases as descriptors are properly reorganized. The reason of this is that the values of P_{pick} and P_{drop} functions, defined in formulas (1) and (3), decrease as descriptors are correctly replaced and sorted on the Grid.

With a mechanism inspired by ants and other insects, each agent maintains a *pheromone base* (a real value) and increases it when its activeness tends to decrease; the agent switches to the *move* mode as soon as the pheromone level exceeds a defined threshold T_h. In particular, at given time intervals, i.e. every 2000 seconds, each agent counts up the number of times that it has evaluated the pick and drop probability functions, $N_{attempts}$, and the number of times that it has actually performed pick and drop operations, $N_{operations}$. At the end of each time interval, the agent makes a deposit into its pheromone base, which is inversely proportional to the fraction of performed operations. An evaporation mechanism is used to give a greater weight to the recent behavior of the agent. Specifically, at the end of the i-th time interval, the pheromone level Φ_i is computed with formulas (4) and (5).

$$\Phi_i = Ev \cdot \Phi_{i-1} + \phi_i \qquad (4)$$

$$\phi_i = 1 - \frac{N_{operations}}{N_{attempts}} \qquad (5)$$

The evaporation rate E_v is set to 0.9 [18], whereas ϕ_i is the amount of pheromone deposited in the last time interval. The pheromone level can assume values comprised between 0 and 10: the superior limit can be obtained by equalizing Φ_i to Φ_{i-1} and setting ϕ_i to 1. As soon as the pheromone level exceeds the threshold T_h (whose value must also be set between 0 and 10), the agent switches its mode from *copy* to *move*. The value of T_h can be used to tune the number of agents that work in the *copy* mode, and consequently the replication and dissemination of descriptors, since these agents are able to generate new descriptor replicas. Specifically, the number of agents in *copy* increases with the value of the threshold T_h. This phenomenon is not analyzed here but is widely discussed in [11].

3.4   Management of Peer Disconnections

In a dynamic Grid, peers can go down and reconnect again with varying frequencies. To account for this, we define the *average connection time* of a peer, which is generated according to a Gamma probability function, with an average value set to the parameter T_{peer}. Use of the Gamma distribution assures that the Grid contains very dynamic hosts, which frequently disconnect and rejoin the network, as well as much more stable hosts.

As a consequence of this dynamic nature, two issues are to be tackled. The first is related to the management of new resources provided by new or reconnected hosts. Indeed, if all the replication agents switch to the *move* mode, it becomes impossible to replicate and disseminate descriptors of new resources; as a consequence, agents cannot be allowed to live forever, and must gradually be replaced by new agents that set off in the *copy* mode. The second issue is that the system must remove "obsolete descriptors", i.e. descriptors of resources provided by hosts that have left the system, and therefore are no longer available.

Simple mechanisms are adopted to cope with these two issues. The first is to correlate the lifecycle of agents to the lifecycle of peers. When joining the Grid, a host generates a number of agents given by a discrete Gamma stochastic function, with average N_{gen}, and sets the life-time of these new agents to the average connection time of the peer itself. This setting assures that (i) the relation between the number of peers and the number of agents is maintained with time (more specifically, the overall number of agents is approximately equal to the number of active peers times N_{gen}) and (ii) a proper turnover of agents is achieved, which allows for the dissemination of descriptors of new resources, since new agents start in the *copy* mode. A second mechanism assures that, every time a peer disconnects from the Grid, it loses all the descriptors previously deposited by agents, thus contributing to the removal of obsolete descriptors. Finally, a *soft state* mechanism [20] is adopted to avoid the accumulation of obsolete descriptors in very stable nodes. Each host periodically refreshes the descriptors corresponding to the resources owned by other hosts, by contacting these hosts and retrieving from them updated information about resources.

It is worth mentioning that the described approach for handling a dynamic Grid implicitly manages any unexpected peer fault, because this occurrence is processed in exactly the same way as a peer disconnection. Indeed, the two events are indistinguishable, since (i) a peer does not have to perform any procedure before leaving the system, and (ii) in both cases (disconnection and fault) the descriptors that the peer has accumulated so far are removed.

4   Performance Evaluation

The performance of the Antares algorithm was evaluated with an event-based simulator written in Java. Simulation objects are used to emulate Grid peers and Antares agents. Each object reacts to external events according to a finite state automaton and responds by performing specific operations and/or by generating new messages/events to be delivered to other objects.

A Grid network having a number of hosts equal to N_p is considered in this work. Hosts are linked through P2P interconnections, and each host is connected to 4 peer hosts on average. The topology of the network was built using the well-known scale-free algorithm defined by Albert and Barabasi [3], that incorporates the characteristic of preferential attachment (the more connected a node is, the more likely it is to receive new links) that was proved to exist widely in real networks. The average connection time of a peer (see Section 3.4) is set to T_{peer}. The number of Grid resources owned and published by a single peer is obtained with a Gamma stochastic function with an average value equal to 15 (see [15]).

Resources are characterized by metadata descriptors indexed by bit strings (keys) having 4 bits, with $2^4 - 1$ possible values [2]. These values, as mentioned in the Introduction, can result from a semantic description of a resource, in which each bit represents the presence of a particular topic, or from the application of a locality preserving hash function. In any case, it is guaranteed that similar keys are given to descriptors of similar resources. Notice that if the range of possible resource types is larger than $2^4 - 1$, it can always be assumed that a resource is characterized by several attributes, each having no more than $2^4 - 1$ possible values. In this case, each attribute corresponds to a separate mapping on the Grid network. This assumption is made in several P2P architectures [2] [5]; however, here we discuss the simple case of one attribute descriptors.

The mean number of agents generated by a single peer is set to the parameter N_{gen}. The average number of agents N_a that travel the Grid is approximately equal to $N_p \cdot N_{gen}$, as explained in Section 3.4. The average time T_{mov} between two successive agent movements is set to 60 s, whereas the maximum number of P2P hops that are performed within a single agent movement, H_{max}, is set to 3, in order to limit the traffic generated by agents. The threshold T_h is set to 9.0, which means that each agent sets off in the *copy* mode and passes to the *move* mode as soon as its pheromone, starting from 0, exceeds the threshold of 9.0 (see Section 3.3).

The parameters N_p, N_{gen} and T_{peer} were set to different values in different sets of experiments, as discussed later. However, their default values were fixed, respectively, at 2500 hosts, 0.5 and 100000 seconds, unless otherwise stated.

Prior to the numerical analysis, a graphical description of the behavior of the algorithm, for the case in which the number of bits in the descriptor key is set to 3, is given in Figure 2. For the sake of clearness we show a portion of the Grid, and Grid hosts are arranged in a bi-dimensional mesh instead of a scale-free network. Each peer is visualized through a color that results from the application of the RGB color model. This color results from a combination of the 3 primary colors (red, green and blue), each of which is associated to one of the three bits of the descriptor (for example red is associated to the first bit and so on) and can assume a value ranging from 0 to 255. For a given peer, the value of each primary color is set by examining the corresponding bits of the descriptors maintained in this peer. More specifically, it is set to the fraction of bits equal to 1 with respect to the total number of descriptors, and then multiplied by 255. This way the color of each peer immediately represents the descriptors that this peer maintains. Three snapshots of the network are depicted: the first is taken when the Antares process is initiated (time 0), the second is taken 50000 seconds later, and the third snapshot is taken in a quite steady situation, 500000 seconds after the process start. This figure shows that descriptors are initially distributed in a completely random fashion, but subsequently they are reorganized and spatially sorted by agents. In fact we note the creation of color spots that reveal the accumulation of similar descriptors in restricted regions of the Grid. The different colors of

---

[2] A key string with all bits equal to 0 is not permitted in order to correctly calculate the cosine of the angle between two vectors.

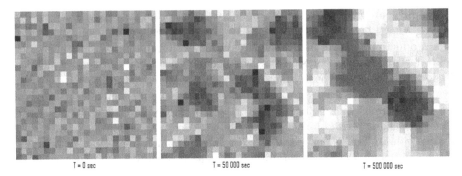

T = 0 sec   T = 50 000 sec   T = 500 000 sec

Fig. 2. Accumulation and reorganization of resource descriptors

the spots correspond to the different combinations of descriptor values that are aggregated. For example red, green and blue spots reveal a massive presence of descriptor keys equal to [1,0,0], [0,1,0], [0,0,1], respectively, while yellow, cyan, violet and white spots represent the predominant presence of vectors having two or three bits equal to 1. We can also observe that colors change gradually between neighbor regions, which proves that the descriptors are not only clustered but also spatially *sorted* on the network.

A set of performance indices are defined to evaluate the performance of the Antares algorithm. The overall homogeneity function H, discussed in Section 3.3, is used to estimate the effectiveness of the algorithm in the reorganization of descriptors. The N_d index is defined as the mean number of descriptors maintained by a Grid host. Since new descriptors are only generated by agents that work in the *copy* mode, the number of such agents, N_{copy}, is another interesting index that helps understand what happens in the system. Finally, the processing load, L, is defined as the average number of agents per second that get to a Grid host, and there perform pick and drop operations.

Performance indices were obtained by varying several parameters, for example the average number of resources published by a host and the frequency of agent movements. We found that the qualitative behavior of Antares is not affected by these parameters, which proves the robustness of the algorithm. This robustness derives from the decentralized, self-organizing and adaptive features of the algorithm, which are also the cause of its scalability. Indeed, since each agent operates only on the base of local information, performance is not significantly affected by the size of the network, as better discussed later.

In this paper, we choose to show performance indices, versus time, obtained for different values of the parameters T_{peer}, N_{gen} and N_p. In the first set of experiments, Antares was evaluated in networks having different churn rates, i.e., with different values of the average connection time of a peer. Specifically, tested values of T_{peer} ranged from 35000 to 1000000 seconds and, for comparison purposes, the case in which peers never disconnect was also tested. This kind of analysis is valuable because it helps understand the mechanisms through which the information system is constructed, and also because it is possible to assess

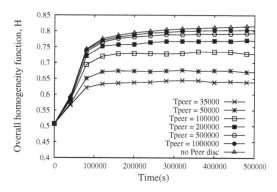

Fig. 3. Overall homogeneity function, vs. time, for different values of the average connection time T_{peer}

the algorithm ability to adapt the mapping of descriptors to the continuous modifications of the environment.

Figure 3 reports the trend of H, the overall homogeneity function. It appears that the work of Antares agents make this index increase from about 0.50 to much higher values. After a transient phase, the value of H becomes stable: it means that the system reaches an equilibrium state despite the fact that peers go down and reconnect, agents die and others are generated, etcetera. In other words, the algorithm adapts to the varying conditions of the network and is robust with respect to them. Note that the stable value of H decreases as the network becomes more dynamic (that is, with lower values of T_{peer}), because the reorganization of descriptors performed by agents is partly hindered by environment modifications. However, even with the lowest value of T_{peer} that we tested, 35000, the percentage increase of H is about 30%, whereas it is more than 60% with T_{peer} equal to 1000000.

Figure 4 depicts N_{copy}, the number of agents that operate in the *copy* mode, also called *copy agents* in the following. This analysis is interesting because *copy* agents are responsible for the replication of descriptors, whereas agents in the *move* mode are exclusively devoted to the relocation and spatial sorting of descriptors.

When the process is initiated, all the agents (about 1250, half the number of peers, since N_{gen} is set to 0.5) are generated in the *copy* mode, but subsequently several agents switch to *move*, as soon as their pheromone value exceeds the threshold T_h. This corresponds to the sudden drop of curves that can be observed in the left part of Figure 4. Thereafter the number of *copy* agents gets stabilized, even with some fluctuations; this equilibrium is reached because the number of new agents which are generated by Grid hosts (these agents set off in the *copy* mode) and the number of agents that switch from *copy* to *move* get balanced.

Figure 4 also shows that the number of *copy* agents increases as the Grid becomes more dynamic. Indeed, a higher turnover of agents is obtained when peers disconnect and reconnect with a higher frequency, because more agents die

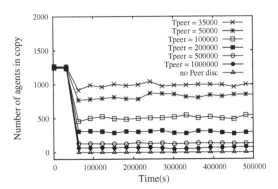

Fig. 4. Number of agents that operate in the *copy* mode, vs. time, for different values of the average connection time T_{peer}

if more peers disconnect (because the lifetime of agents is correlated to the life-time of peers), and at the same time more agents are generated by reconnecting peers (see Section 3.4). Since new agents set off in the *copy* mode, this leads to a larger number of *copy* agents, as appears in Figure 4. Moreover, note that in a stable network (no peer disconnections) all agents work in the *move* mode after a short transient phase. Indeed in this case no new agents are generated after the process is initiated.

Figure 5 reports N_d, the average number of descriptors that are maintained by a Grid host at a given time. One of the main objectives of Antares is the replication and dissemination of descriptors. This objective is achieved because the value of N_d increases from an initial value of about 15 (equal to the average number of resources published by a host) to much higher values; as for the other indices, the trend of N_d undergoes a transient phase, then it becomes stabilized, even if with some fluctuations. The value of N_d is determined by two main phenomena: on the one hand, a large number of *copy* agents tend to increase N_d, because they are responsible for the generation of new descriptor replicas. On the other hand, a frequent disconnection of peers tends to lower N_d, because a disconnecting peer loses all the descriptors that it has accumulated so far (see Section 3.4). These two phenomena work in opposite directions as the value of T_{peer} increases: in a more dynamic network there are more *copy* agents (which tends to increase N_d), but more descriptors are thrown away by disconnecting peers (which tends to decrease N_d).

The result is that the stable number of N_d is relatively lower both when the dis-connection frequency is very low and when it is very high or infinite (i.e., with no peer disconnections). Interestingly, a higher degree of replication can be reached for intermediate values of T_{peer}, which are more realistic on Grids. Indeed, Figure 5 shows, even if curves are more wrinkled than those examined so far (probably due to the two underlying and contrasting mechanisms discussed above), that the value of N_d first increases as T_{peer} increases from 35000 to values comprised be-tween 50000 and 100000, then it decreases again for higher values of T_{peer}.

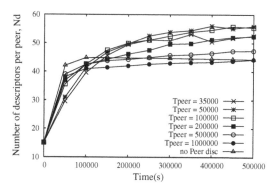

Fig. 5. Average number of descriptors maintained by a Grid host, vs. time, for different values of the average connection time T_{peer}

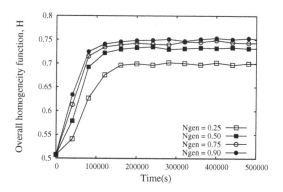

Fig. 6. Overall homogeneity function, vs. time, for different values of N_{gen}

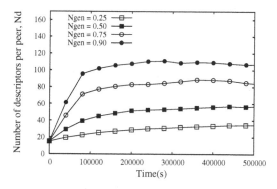

Fig. 7. Average number of descriptors maintained by a Grid host, vs. time, for different values of N_{gen}

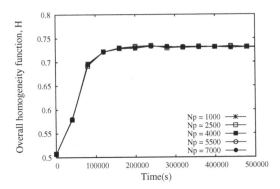

Fig. 8. Overall homogeneity function, vs. time, for different values of the number of peers N_p

Figures 6 and 7 show the values of the homogeneity function and the average number of descriptors stored by a host, for different values of N_{gen}, the average number of agents generated by a reconnecting host. It is very interesting to note that the value of N_{gen} can be used to tune the replication of descriptors. In fact, since the number of agents is proportional to the value of N_{gen} (see Section 3.4), the steady value of N_d increases with the value of N_{gen}, as Figure 7 confirms. Therefore, the value of N_{gen} can be increased to foster the replication of descriptors (for example, because this can facilitate discovery operations) or decreased to reduce the number of descriptors stored on hosts (this can be necessary if the amount of memory available on hosts is limited). On the other hand the value of N_{gen} has a much smaller effect on the spatial reorganization of descriptors, as is testified by Figure 6. This figure shows that the steady value of H is scarcely modified by varying the number of agents that circulate in the network.

The scalability properties of Antares were analyzed by varying the number of peers N_p, from 1000 to 7000. Interestingly, the size of the network has no detectable effect on the performance, specifically on the overall homogeneity index, as shown in Figure 8. This confirms the scalable nature of Antares, which derives from its decentralized and self-organizing characteristics. Similar considerations can be made for other performance indices, such as the number of descriptors per host, N_d, which is not shown here.

Finally, the processing load L, defined as the average number of agents per second that are processed by a peer, does not depend neither on the churn rate nor on the network size, but only depends on the number of agents and the frequency of their movements across the Grid. L can be obtained as follows:

$$L = \frac{N_a}{N_p \cdot T_{mov}} = \frac{N_{gen}}{T_{mov}} \tag{6}$$

In the reference scenario, the average value of T_{mov} is equal to 60 seconds, and the default value of N_{gen} is fixed at 0.5. Therefore, each peer receives and processes about one agent every 120 seconds, which can be considered an acceptable load.

In the Introduction of the paper, we briefly mentioned that the reorganization and sorting of descriptors can be exploited by a discovery algorithm that allows users to find the resources that they need for their applications. A query can be issued to search for "target descriptors", that is, for resource descriptors having a given value of their binary index. Thanks to the spatial sorting of descriptors achieved by ant-based agents, the discovery procedure can be simply managed by forwarding the query, at each step, towards the "best neighbor", that is, the neighbor peer that maximizes the similarity between the descriptors stored locally and the target descriptor. Preliminary experiments are confirming that this approach allows query to get to a large number of useful descriptors in most cases.

5   Conclusions

In this paper we introduced and evaluated Antares, an algorithm inspired on the behavior of ants in a colony, whose aim is to build a P2P information system of a Grid. Through the evaluation of simple probability functions (*pick* and *drop*), a number of ant-inspired agents replicate and move the descriptors of Grid resources from host to host, and this way disseminate and reorganize these descriptors on the network.

Antares achieves an effective reorganization of information, since descriptors are spatially sorted on the network and, in particular, descriptors indexed by equal or similar binary keys are placed in neighbor Grid hosts. This was confirmed in the paper both with a graphical description based on the RGB model and with the analysis of performance measures, in particular of a homogeneity index based on the cosine similarity between binary vectors.

Antares is scalable and robust with respect to the variation of algorithm and network parameters. In particular, the reorganization of descriptors performed by Antares spontaneously adapts to the ever changing environment, for example to the joins and departs of Grid hosts and to the changing characteristics of resources. This results from the decentralized, self-organizing and adaptive features of Antares, which are borrowed by the corresponding biological system.

The resulting P2P information system is basically *unstructured* because there is no predetermined association between resources and hosts, but thanks to the work of agents, the spatial sorting of descriptors allows important benefits of *structured* P2P systems to be retained. Specifically, the reorganization of descriptors enables the possibility of effectively serving both simple and *range* queries, since descriptors indexed by similar keys are likely to be located in neighbor hosts. The confirmation of this intuition is one of the main objectives of current work.

Acknowledgments. This research work is carried out under the FP6 Network of Excellence CoreGRID funded by the European Commission (Contract IST-2002-004265). This work has also been supported by the Italian MIUR FAR SFIDA project on interoperability in innovative e-business models for SMEs through an enabling Grid platform.

References

1. Ghosh, S., Briggs, R., Padmanabhan, A., Wang, S.: A self-organized grouping (sog) method for efficient grid resource discovery. In: Proc. of the 6th IEEE/ACM International Workshop on Grid Computing, Seattle, Washington, USA (November 2005)
2. Andrzejak, A., Xu, Z.: Scalable, efficient range queries for grid information services. In: Proc. of the Second IEEE International Conference on Peer-to-Peer Computing P2P 2002, Washington, DC, USA, pp. 33–40. IEEE Computer Society, Los Alamitos (2002)
3. Barabási, A.-L., Albert, R.: Emergence of scaling in random networks. Science 286(5439), 509–512 (1999)
4. Bonabeau, E., Dorigo, M., Theraulaz, G.: Swarm intelligence: from natural to artificial systems. Oxford University Press, New York (1999)
5. Cai, M., Frank, M., Chen, J., Szekely, P.: Maan: A multi-attribute addressable network for grid information services. In: GRID 2003: Proceedings of the Fourth International Workshop on Grid Computing, Washington, DC, USA, p. 184. IEEE Computer Society, Los Alamitos (2003)
6. Camazine, S., Franks, N.R., Sneyd, J., Bonabeau, E., Deneubourg, J.-L., Theraula, G.: Self-Organization in Biological Systems. Princeton University Press, Princeton (2001)
7. Chakravarti, A.J., Baumgartner, G., Lauria, M.: The organic grid: self-organizing computation on a peer-to-peer network. IEEE Transactions on Systems, Man, and Cybernetics, Part A 35(3), 373–384 (2005)
8. Crespo, A., Garcia-Molina, H.: Routing indices for peer-to-peer systems. In: Proc. of the 22nd International Conference on Distributed Computing Systems ICDCS 2002, pp. 23–33 (2002)
9. Dorigo, M., Bonabeau, E., Theraulaz, G.: Ant algorithms and stigmergy. Future Generation Compututer Systems 16(9), 851–871 (2000)
10. Erdil, D.C., Lewis, M.J., Abu-Ghazaleh, N.: An adaptive approach to information dissemination in self-organizing grids. In: Proc. of the International Conference on Autonomic and Autonomous Systems ICAS 2006, Silicon Valley, CA, USA (July 2005)
11. Forestiero, A., Mastroianni, C., Spezzano, G.: Construction of a peer-to-peer information system in grids. In: Czap, H., Unland, R., Branki, C., Tianfield, H. (eds.) Self-Organization and Autonomic Informatics (I). Frontiers in Artificial Intelligence and Applications, vol. 135, pp. 220–236. IOS Press, Amsterdam (2005)
12. Forestiero, A., Mastroianni, C., Spezzano, G.: Reorganization and discovery of grid information with epidemic tuning. Future Generation Computer Systems 24(8), 788–797 (2008)
13. Forestiero, A., Mastroianni, C., Spezzano, G.: So-Grid: A self-organizing grid featuring bio-inspired algorithms. ACM Transactions on Autonomous and Adaptive Systems 3(2) (May 2008)
14. Foster, I., Kesselman, C.: The Grid 2: Blueprint for a New Computing Infrastructure. Morgan Kaufmann Publishers Inc, San Francisco (2003)
15. Iamnitchi, A., Foster, I., Weglarz, J., Nabrzyski, J., Schopf, J., Stroinski, M.: A peer-to-peer approach to resource location in grid environments. In: Grid Resource Management. Kluwer Publishing, Dordrecht (2003)
16. Lumer, E.D., Faieta, B.: Diversity and adaptation in populations of clustering ants. In: Proc. of SAB 1994, 3rd international conference on Simulation of adaptive behavior: from animals to animats 3, pp. 501–508. MIT Press, Cambridge (1994)

17. Oppenheimer, D., Albrecht, J., Patterson, D., Vahdat, A.: Design and implementation tradeoffs for wide-area resource discovery. In: Proc. of the 14th IEEE International Symposium on High Performance Distributed Computing HPDC 2005, Research Triangle Park, NC, USA (July 2005)
18. Van Dyke Parunak, H., Brueckner, S., Matthews, R.S., Sauter, J.A.: Pheromone learning for self-organizing agents. IEEE Transactions on Systems, Man, and Cybernetics, Part A 35(3), 316–326 (2005)
19. Platzer, C., Dustdar, S.: A vector space search engine forweb services. In: ECOWS 2005: Proceedings of the Third European Conference on Web Services, Washington, DC, USA, p. 62. IEEE Computer Society, Los Alamitos (2005)
20. Sharma, P., Estrin, D., Floyd, S., Jacobson, V.: Scalable timers for soft state protocols. In: Proc. of the 16th Annual Joint Conference of the IEEE Computer and Communications Societies, INFOCOM 1997, Washington, DC, USA, vol. 1, pp. 222–229. IEEE Computer Society, Los Alamitos (1997)
21. Taylor, I.J.: From P2P to Web Services and Grids: Peers in a Client/Server World. Springer, Heidelberg (2004)

Robustness to Code and Data Deletion in Autocatalytic Quines

Thomas Meyer[1], Daniel Schreckling[2], Christian Tschudin[1],
and Lidia Yamamoto[1]

[1] Computer Science Department, University of Basel
Bernoullistrasse 16, CH–4056 Basel, Switzerland
{th.meyer,christian.tschudin,lidia.yamamoto}@unibas.ch
[2] Computer Science Department, University of Hamburg
Vogt-Koelln-Str. 30, D–22527 Hamburg, Germany
schreckling@informatik.uni-hamburg.de

Abstract. Software systems nowadays are becoming increasingly complex and vulnerable to all sorts of failures and attacks. There is a rising need for robust self-repairing systems able to restore full functionality in the face of internal and external perturbations, including those that affect their own code base. However, it is difficult to achieve code self-repair with conventional programming models.

We propose and demonstrate a solution to this problem based on self-replicating programs in an artificial chemistry. In this model, execution proceeds by chemical reactions that modify virtual molecules carrying code and data. Self-repair is achieved by what we call *autocatalytic quines*: programs that permanently reproduce their own code base. The *concentration* of instructions reflects the health of the system, and is kept stable by the instructions themselves. We show how the chemistry of such programs enables them to withstand arbitrary amounts of random code and data deletion, without affecting the results of their computations.

1   Introduction

Today's computer programs are engineered by humans using the classical technique of building an abstract model of the real world and implementing reactions to this modelled world. Hence it is not surprising that these programs often fail in the real world environment, because their designers did not foresee certain situations, because of implementation errors or malicious attacks. In addition, as devices become smaller the influence of electical noise and cosmic radiation increases, resulting in transient faults that lower the reliablity of the calculation outcome [1]. Thus there is a need to build reliable systems based on unreliable components where the software is resilient against internal and external perturbations and recovers itself from such situations.

Self-recovery means adapting the internal model to maintain properties like correctness, effectiveness, etc. This generic kind of adaptation is not possible with pre-programmed strategies. Because most programming languages available

C. Priami et al. (Eds.): Trans. on Comput. Syst. Biol. X, LNBI 5410, pp. 20–40, 2008.
© Springer-Verlag Berlin Heidelberg 2008

today separate code from data, a program once deployed cannot be changed in order to adapt to environmental changes. This lack of elasticity makes it hard to build programs that react to unpredicted situations. One approach to address the challenge of constructing self-healing software could be the use of self-modifying programs. However, there is still little theory about how to design such programs [2].

Our approach is to use self-replicating programs in an artificial chemistry. Artificial chemical computing models [3,4,5] express computations as chemical reactions that consume and produce data or code objects. In an artificial chemistry, computations can occur at microscopic and macroscopic scales [4]. At the microscopic scale, we observe how reactions operate on the information stored in individual molecule instances; at this scale, a change in a single molecule has a visible and immediate impact on the result of the computation. At the macroscopic scale, in contrast, molecules occur at massive numbers, and a change in a single molecule is unlikely to have a significant impact on the system as a whole; at this level, it is the concentration of different molecular species that mainly determines the outcome of the computation, which is ready when the system reaches a steady state [6,7].

While artificial chemistries designed for biological modelling have focused on computation at macroscopic scales [7,8], chemical computing abstractions targeted at concurrent information processing have traditionally focused on microscopic scales [9,10]. We adopt a hybrid approach in which code molecules are able to self-replicate in order to maintain their concentration and survive deletion attacks. This results in a fault tolerance mechanism observable at the macroscopic scale, while computations are performed at the microscopic level. These self-replicating code molecules are called *autocatalytic quines*. Quines are programs that produce their own code as output, and autocatalytic refers to their ability to make copies of themselves, thus catalyse their own production.

We show how autocatalytic quines are able to perform some simple function computation while at the same time keeping their concentrations stable in the presence of a dilution flux, and also in the presence of targeted deletion attacks with a significant decrease in their molecule concentrations. The resulting fault tolerance is based on a constantly dynamic and regenerating system, as opposed to traditional techniques which seek to maintain the system as static as possible. All our experiments are performed using the *Fraglet* programming language [11,12] in which self-replicating code can be programmed in a straightforward way.

This paper is structured as follows: Section 2 introduces self-replicating systems and related chemical models. Section 3 briefly explains the Fraglet model and instruction set. Section 4 shows how such self-replicating sets of "molecules" can be built in Fraglets, from simple quines up to generic operations. Section 5 then analyzes the dynamic aspects of self-replicating sets. We show how to achieve robustness to deletion of molecules by imposing a dilution flow to the reaction vessel that limits the growth of autocatalytic quines. In Sect. 6 we show a robust program that calculates an arithmetic expression using quine operations.

Finally, in Sect. 7 the characteristics of Fraglets and their reaction model will serve as a case study for a brief security analysis of artificial chemical computing models.

2 Background and Related Work

The search for potential models of machines that can produce copies of themselves can be traced back to the late 1940's, with the pioneering work by John von Neumann on a theory of self-reproducing automata [13]. He described a universal constructor, a (mechanical) machine able to produce a copy of any other machine whose description is provided as input, including a copy of itself, when fed with its own description. Both the machine and the description are copied in the process, leading to a new machine that is also able to replicate in the same way.

The definition in [14] makes the distinction between replication and reproduction clear: Replication involves no variation mechanism, resulting in an exact duplicate of the parent entity; deviations from the original are regarded as errors. On the other hand, reproduction requires some form of variation, for instance in the form of genetic operators such as mutation and crossover, which may ultimately lead to improvement and evolution. These operators change the description of the machine to be copied, requiring a self-modification mechanism.

Replication, reproduction and variation in living beings are performed as chemical processes in the DNA. In the computer science context, they map therefore well to artificial chemical computing models, which attempt to mimic such processes in a simplified way. Numerous such artificial chemistries have been proposed [5], with the most varied purposes from studying the origins of life to modelling chemical pathways in cells, or simply providing inspiration for new, highly decentralized computing models.

In this section we discuss related research in self-replicating code and artificial chemistries.

2.1 Self-replicating Code

Since von Neumann set the basis for a mathematically rigorous study of self-replicating machines, many instances of such machines have been proposed and elaborated. An overview of the lineage of work in the area of self-replication can be found in [15,16].

Self-replication is a special case of universal construction, where the input to the constructor (description) contains a description of itself. However, while universal construction is a sufficient condition for self-replication, it is not a necessary one. Indeed Langton [17] argued that natural systems are not equipped with a universal constructor. He relaxed the requirement that self-replicating structures must treat their stored information both as interpreted instructions and uninterpreted data. With this he showed that simple self-replicating structures based on dynamic loops instead of static tapes can be built. This spawned a new surge of research on such self-replicating structures [18,16].

Most of the contributions to self-replication were done within the cellular automata (CA) framework, introduced by von Neumann. Self-replicating code was a later branch appearing in the 1960's, focusing on replication of textual computer programs. The work on self-replicating code was motivated by the desire to understand the fundamental information-processing principles and algorithms involved in self-replication, even independent of their physical realization.

The existence of self-replicating programs is a consequence of Kleene's second recursion theorem [19], which states, that for any program p there exists a program p', which generates its own encoding and passes it to p along with the original input. The simplest form of a self-replicating program is a *quine*, named after the philosopher and logician Willard van Orman Quine (1908-2000). A quine is a program that prints its own code. Quines exist for any programming language that is Turing complete. The *Quine Page* [20] provides a comprehensive list of such programs in various languages.

2.2   Artificial Chemistries

Artificial chemical computing models [10,3,4,9] express computations as chemical reactions that consume and produce objects (data or code). Objects are represented as elements in a *multiset*, an unordered set within which elements may occur more than once.

In [5] chemical computing models are classified as applications of Artificial Chemistry, a branch of Artificial Life (ALife) dedicated to the study of the chemical processes related to life and organizations in general. In the same way as ALife seeks to understand life by building artificial systems with simplified life-like properties, Artificial Chemistry builds simplified abstract chemical models that nevertheless exhibit properties that may lead to emergent phenomena, such as the spontaneous organization of molecules into self-maintaining structures [21,22]. The applications of artificial chemistries reach biology, information processing (in the form of natural and artificial chemical computing models) and evolutionary algorithms for optimization, among other domains.

Chemical models have also been used to express replication, reproduction and variation mechanisms [23,5,24,25]. From these, we focus on models that apply these mechanisms to computer programs expressed in a chemical language, especially when these programs are represented as molecular chains of atoms that can operate on other molecular species, as opposed to models where a finite and well-known number of species interact according to predefined, static reaction rules. The interest of such molecular-chain models is two fold: first of all, complex computations can be expressed within molecule chains; second, they can more easily mimic the way in which DNA, RNA and enzymes direct reproduction, potentially leading to evolution in the long run.

Holland's Broadcast language [26] was one of the very earliest computing models resembling chemistry. It also had a unified code and data representation, in which broadcast units represented condition-action rules and signals for other units. These units also had self-replication capacity, and the ability to detect the presence or absence of a given signal in the environment. The language was recently implemented [27], and revealed helpful in modelling real biochemical signalling networks.

In [5], several so-called artificial polymer chemistries are described. In these systems, molecules are virtual polymers, long chains of "monomers" usually represented as letters. Polymers may concatenate with each other or suffer a cleavage at a given position. The focus of those models was to model real chemistries or to study the origin of life. In [23] pairs of simple fixed-length binary strings react with each other: one of them represents the code and the other the data on which the code operates. The authors show the remarkable spontaneous emergence of a crossover operator after some generations of evolutionary runs. More recently [28], a chemistry based on two-dimensional molecular chains (represented as strings of lines) has been proposed to model molecular computing, and has been shown to be able to emulate Turing machines.

We conjecture that a chemical language can express programs that can be more easily transformed and can become more robust to disruptions due to alternative execution paths enabled by a multiset model. Therefore they lend themselves more easily to self-modification, replication and reproduction. However, there are difficulties: the non-deterministic, decentralized and self-organizing nature of the computation model make it difficult for humans to control such chemical programs.

3   The Fraglet Reaction Model

Fraglet is an execution model inspired by chemistry originally aimed at the synthesis and evolution of communication protocols [11]. A fraglet, or computation fragment, is a string of atoms (or symbols) $[s_1 \ s_2 \ s_3 \ \ldots \ s_n]$ that can be interpreted as a code/data sequence, as a virtual "molecule" used in a "chemical reaction", or as a sequence of packet headers, or yet as an execution thread. There are a fixed number of production rules describing substitution pattern that operate on fraglets. We limit ourselves to substitution patterns which, on their left side, only depend on the first symbol of a word. For example, the rule [exch S T U TAIL] $\rightarrow$ [S U T TAIL] when applied to the word [exch a b c d] will result in [a c b d] – that is, two symbols are swapped. The exch acted as a prefix command for the rest of the word whereas the new left-most symbol 'a' serves as a continuation pointer for further processing of the result. Table 1 shows some of the production rules. More details on the instruction set can be found in [11] with an update in [12].

According to [5], an artificial chemistry can be broadly defined by a triple (S, R, A), where S is the set of molecular species, R is the set of collision (reaction) rules, and A is the algorithm for the reaction vessel where the molecules S interact via the rules R. In the case of Fraglets, the set of molecules S consists of all possible fraglets. (The term molecule and fraglet is used synonymously in this paper.) The finite set of production rules implicitly define the potentially infinite set of reactions R among molecules.

The dynamic behaviour of a simulation is characterized by the algorithm A. Fraglets are injected into a virtual reaction vessel, which maintains a multiset of

Table 1. Selected production rules of a Fraglet system. `S`, `T` and `U` are placeholders for symbols, `TAIL` stands for a potentially empty word of symbols.

Instruction	Educt(s)	Product(s)
exch	[exch S T U TAIL] $\to$ [S U T TAIL]	
send	n_i[send n_j TAIL] $\to$ $_{n_j}$[TAIL]	
split	[split PART1 * PART2] $\to$ [PART1] + [PART2]	
fork	[fork S T TAIL] $\to$ [S TAIL] + [T TAIL]	
sum	[sum S i_1 i_2 TAIL] $\to$ [S i_1+i_2 TAIL] (do. for mult etc)	
match	[match S TAIL1] + [S TAIL2] $\to$ [TAIL1 TAIL2]	
matchp	[matchp S TAIL1] + [S TAIL2] $\to$ [matchp S TAIL1] + [TAIL1 TAIL2]	

fraglets and simulates their reactions using the Gillespie algorithm [29]. The algorithm calculates the collision probability of two molecules in a well-stirred tank reactor. The reaction rate is proportional to the product of the concentrations of the reaction educts.

Fraglet programs normally consist of a set of active `match` or `matchp` fraglets that process passive (data) molecules. In order to make such programs robust to the deletion of constituting parts we aim at constantly replicate the `match` fraglets while they are being executed. In the next sections we show how self-replication can be realized using the Fraglet reaction model. In Sect. 4 we examine how to replicate the information stored in a set of fraglets. To analyze these static aspects of self-replication we focus on the set of molecules and reaction rules (S, R). Then, in Sect. 5 we include the dynamic behavior into our considerations and therefore discuss the impact of the reaction algorithm A in more detail.

4   Self-replication in Fralgets

In this section we provide examples of self-replication in the Fraglet framework. Our goal is to find a set of fraglets that is able to maintain itself, i.e. to replicate all fraglets that are part of this set. We use quines as templates for self-replicating programs, and show how to embed useful functions into them.

In general, a quine consists of two parts, one which contains the executable code, and the other which contains the data. The data represents the blueprint of the code. The information that is stored in the blueprint is used twice during replication: First it serves as instructions to be interpreted by the quine to construct a new quine. Then the same information is attached to the new offspring, so that it is able to replicate in turn.

The usage of information to build instructions can be compared to the *translation* of genes occurring in cells, where RNA chains carrying the genotype are translated into proteins. The latter use of the blueprint resembles the DNA *replication*.

In Fraglets it is easy to implement both *translation* and *replication* of data molecules, since the Fraglet language has a flat code/data representation. The following example shows a reaction trace of a data fraglet **bp** that is *translated* (interpreted as executable code) by a **match** rule.

[match bp] + [bp sum y 4 5] $\longrightarrow$ [sum y 4 5] $\longrightarrow$ [y 9]

This *translation* process in Fraglets is performed as an explicit removal of the passive header tag which activates the rule such that it can be executed. The "collision" of the active **match** and the passive fraglet cause the active and the passive parts react together, resulting in a new active fraglet that calculates the sum. Information *replication*, on the other hand, can be achieved by using a fork instruction. Hence a simple quine can be built by finding a code and a data fraglet that react and, in doing so, regenerate both.

Here is an example of a simple self-replicating, autocatalytic quine program:

```
[   match bp fork fork fork nop bp]
[bp match bp fork fork fork nop bp]
```

In this example, depicted in Fig. 1, two copies of the information are present: the first is the executable (active) copy, and the second is the code storage (blueprint), guarded by tag **bp**. Their reaction produces a fraglet starting with [**fork fork** ...] which *replicates* the information by generating two copies of the remaining part. The resulting two identical fraglets again start with a **fork** instruction. They individually *translate* the information: One copy reduces to the active part, restarting the cycle, and the other reinstalls the original blueprint.

We observe that the active and the passive parts look similar. In fact, they only differ in their head symbol: the passive part is tagged with a tag **bp**, whereas the active part directly starts with the **match** instruction. Note that a single "seed" fraglet is sufficient to generate (bootstrap) the quine.

4.1   Embedding Functionality

The quine presented before is an example for simple self-replication. It does not do any useful computation and spends cycles only to replicate itself. Some

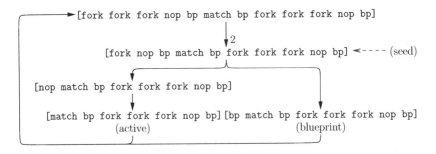

Fig. 1. Simple autocatalytic quine

related work also concentrate on the structure of such self-replicating sets of molecules: Kaufmann, Bagley, Farmer, and later Mossel and Steel examined the properties of random autocatalytic sets [30,31,32,33], Fontana [22] as well as Speroni di Fenizio [34] studied the formation of organizations in a random reaction vessel of λ-expressions and combinator algebra, respectively. Most of them concentrate on the structure of self-replicating sets. Here we aim at performing microscopic computation [4], where the computation is carried out on instances of input molecules and the result is stored to an instance of an output molecule. A simple example for such microscopic computation is the following Fraglet reaction that computes the expression $y = x + 1$:

```
[match x sum y 1] + [x 5] ⟶ [sum y 1 5] ⟶ [y 6]
```

where the active match molecule reacts with a passive (data) molecule x that carries the input data. The resulting fraglet sum increments the number stored within itself and immediately produces a passive output molecule y containing the result.

Consider the following quine template, which can compute any function expressed as a *consume* and *produce* part:

[*spawn consume replicate produce*]

where generally

spawn ::= fork nop bp
replicate ::= split match bp fork fork *spawn* *

For example, to generate a quine replacement for [match x sum y 1], we can define

consume ::= match x
produce ::= sum y 1

which results in the following quine *seed* code:

```
[fork nop bp match x split match bp
fork fork fork nop bp * sum y 1]
```

The reaction graph for this quine is shown in Fig. 2. We can distinguish three interdependent pathways: the upper cycle replicates the blueprint, the cycle in the middle generates the active rule ([match x ...]), and the lower pathway consumes an input molecule x, performs the actual computation and produces an output molecule y. Such as for the simple quine, the information of the *decorated quine* itself is copied during execution: one copy becomes the passive form [bp match x ...] and the other is executed. However, the decorated quine can

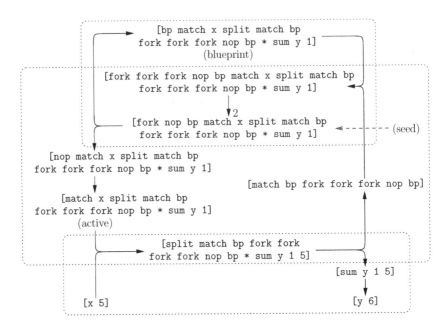

Fig. 2. Decorated quine: Autocatalytic quine with embedded functionality

only replicate when an input molecule x is available. At the same time, the program performs its intended functionality, i.e. to calculate the expression. Quines that perform more complex computations can be written by just specifying the production and consumption sides accordingly.

Operation. The decorated quine is an ideal building block to create larger programs. In this paper we also refer to such a building block as an *operation*. The graphical short notation for the operation that calculates $y = x + 1$ is depicted in Fig. 3. It hides the replication details and only reveals the performed calculation.

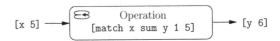

Fig. 3. Operation: Short notation for a decorated quine

5   Dynamic Aspects of Self-replication

In this section we examine the dynamic aspects of the Fraglet algorithm in more detail. First we show two dynamic counterforces, growth and dilution flow that when acting together exhibit the desired property of robustness to code and data deletion.

5.1   Growth

So far we only analyzed the reaction network of the quines from the functional point of view. Let us turn back to the simple quine in Fig. 1 and analyze its dynamic behavior. In each round, the active and the passive part of the quine react and generate two copies of themselves. Each of the two resulting quine instances again reacts and yield two other instances. In a real chemical reaction vessel with constant volume, the collision probability of the reaction partners would rise. Such behavior can be simulated using the Gillespie algorithm [29], which makes sure that the simulated (virtual) time evolves by smaller increments the more possible reactions can take place in parallel. Since the possible reaction partners grow with the number of quine instances, the virtual time increments continuously shrink. As a result the quine population grows exponentially with respect to virtual time: the quine catalyzes its own replication.

For the decorated quine (operation) shown in Fig. 2 we already noted that it only replicates itself when consuming an input molecule x. Thus the replication rate of an operation depends on the rate at which input molecules are injected into the reaction vessel. When input molecules appear at a constant interval, the operation exhibits a linear growth.

5.2   Dilution

The simulation of a growing population of quines on a computer quickly leads to a situation where the virtual time increment of a simulated reaction is smaller than the real time that is needed by the computer to perform the reaction and to calculate the next iteration of the algorithm. In this case the computer cannot perform a real-time simulation of the reaction system anymore which is crucial for programs interfacing a real world environment like programs for robot control or network protocols.

Thus we limit the maximum number of molecules in the reaction vessel to a certain maximum vessel capacity N. For this purpose we apply a random excess dilution flow, which is (1) non-selective, i.e. it randomly picks one of the molecule instances for dilution using a uniform probability distribution, and where (2) the rate of the dilution flow depends on the overall production rate of the system.

Figure 4(a) shows the operating principle of the excess dilution flow. The reaction vessel is in either of two states: in the transient state the number of molecules is smaller than the specified capacity N and the dilution flow is inactive, whereas in the saturation state the number of molecules is equal to the specified limit and the dilution flow is active. Whenever a reaction produces new molecules in the saturation state, the excess dilution flow randomly selects and destroys molecules until the vessel reaches the max. fill level N. The excess dilution flow is able to keep the overall number of molecules at this level even though the population of a molecule set like the quine is growing exponentially.

Formally, the system can be described by the *network replicator equation* [35]. In the next section we will see that the dilution flow acts as a selection pressure mechanism, favoring those quines that are able to self-replicate. A similar method has already been used by others, for example in [22,34].

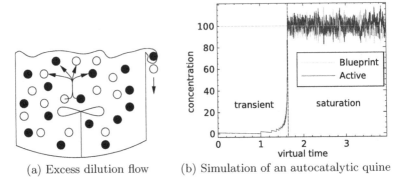

(a) Excess dilution flow   (b) Simulation of an autocatalytic quine

Fig. 4. Excess dilution flow and the time evolution of a simulation of an autocatalytic (exponentially growing) quine using the Gillespie algorithm with a max. vessel capacity of $N = 200$ molecules.

Figure 4(b) shows the dynamic behavior of a simple autocatalytic quine in a reaction vessel limited by an excess dilution flow. In the transient state the quine population grows exponentially with respect to virtual time. When reaching the specified vessel capacity N each newly produced molecule replaces another randomly picked molecule.

The random dilution flow may temporarily favor either the blueprint or the active fraglet of the quine. However, on the long run, this will be compensated, because the dilution mechanism more likely destroys fraglets with a higher concentration, resulting in an equal concentration of $\frac{N}{2} = 100$.

5.3   Emergent Robustness to Code Deletion

One of the emergent properties of an exponentially growing population limited in a finite environment is robustness to the loss of individuals. An unexpected loss of any quine molecule instance decreases the total number of molecules, but this "hole" is immediately filled by the offsprings of the remaining quine instances.

Figure 5 depicts the result of a data and a code deletion attack. At virtual time $t_v = 3$ we remove 80 % of the blueprints. As the replication of the remaining autocatalytic quines continues they quickly replenish the missing molecules. Immediately after the attack the reaction vessel is in the transient state again: The remaining quine instances are able to replicate without being diluted, and therefore the population grows exponentially. When reentering the saturation state, there are temporarily more active rules than blueprints. But the restored dilution flow causes the active fraglets being diluted more frequently. As soon as the equilibrium between blueprints and active rules is reobtained the system reaches its original steady-state and continues operating as before the attack. The same holds for an attack to the active fraglets, shown in Fig. 5 at $t_v = 4$.

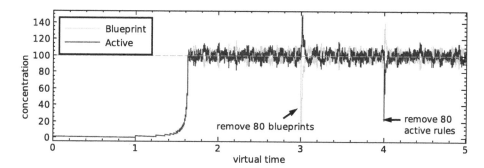

Fig. 5. Code and data deletion attacks. Time evolution of the simulation of an auto-catalytic quine with deletion attacks at $t_v = 3$ (blueprints) and $t_v = 4$ (active fraglets) using the Gillespie algorithm with a max. vessel capacity of $N = 200$ molecules.

6 Robust Programs Using Self-replicating Operations

In this section we focus on how to build programs out of autocatalytic quines (operations). The goal is to arbitrarily combine "quine-protected operations" such that the overall resulting program is still robust to code and data deletions.

6.1 Example: Arithmetic Expression

As an example we build a program that calculates the arithmetic expression $y = 3x^2 + 2x + 1$. Since Fraglets only has binary (arity 2) arithmetic operators we have to split up the calculation into 7 different operations. Each operation consumes a certain input molecule and produces an intermediate result molecule that is then consumed by another operation. Figure 6 shows the reaction flow of the resulting program. The first operation forks the input fraglet x into two intermediate data fraglets t1a, and t1b, respectively. Then further calculation is performed in parallel by two chains. The left chain calculates the square root of the input and multiplies it by 3 whereas the right chain calculates the partial sum of $2x + 1$. Finally, the last instruction joins the resulting data fraglets of the two chains and calculates the sum over the partial results.

Figure 7 shows the result of a simulation. The system quickly recovers from a deletion attack where 80 % of all molecules are removed and returns to its equilibrium where the N molecules are evenly distributed among blueprints and active fraglets of all operations.

6.2 Discussion

We showed how programs can be built by sequentially assembling decorated autocatalytic quines. The simulated burst deletion attacks already gave an indication that the overall system is robust to the deletion of molecules. Here we will further examine the dynamic properties of the resulting system.

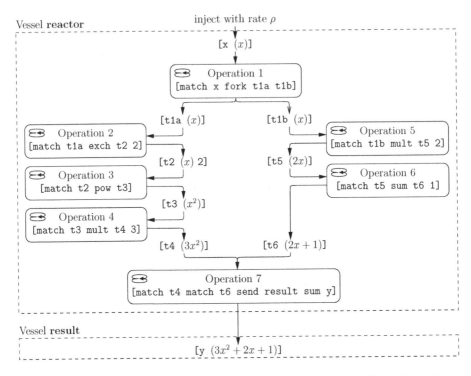

Fig. 6. Parallel chains of catalytic quines (operations) calculating the arithmetic expression $y = 3x^2 + 2x + 1$

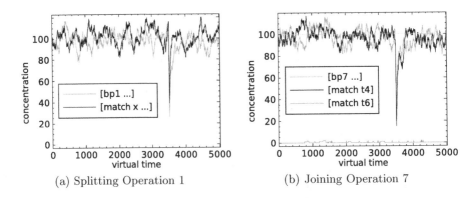

(a) Splitting Operation 1   (b) Joining Operation 7

Fig. 7. Time evolution of the simulation of a partially parallel program to calculate the arithmetic expression $y = 3x^2 + 2x + 1$ using autocatalytic quines (operations). The Gillespie algorithm drives the system, and an excess dilution flow with a maximum vessel capacity of $N = 1400$ restricts unlimited growth. One input molecules x is injected with rate $\rho = 1$. At $t_v = 3500$, 80 % of all molecules in the reaction vessel are destroyed.

Qualitative Considerations. A program is only able to survive in the reaction vessel if all its operations process the same data stream or if all processed data stream rates are in the same order of magnitude: We already observed that the rate of the data stream determines the replication rate of an operation. Because of the dilution flow and the operations competing for concentration, the replication rate of an operation also defines the steady-state concentration of its blueprint and active fraglets. For example, if the replication rate of one operation is twice the replication rate of another, the latter one will be present only with half of the concentration of the first and thus, the probability rises that it will become extinct because of random fluctuations of the algorithm. In our example all operations replicate with more or less the same rate, because all operations process the same data stream (without loops) in one of the parallel branches.

To learn more about the dynamic aspects and characteristics of the overall program we now further analyze the system by changing its input parameters and examining the resulting behavior.

Three input parameters have an impact on the behavior of a given program: (P.1) The max. vessel capacity N is held constant during the simulation. (P.2) The input molecules are actually generated by an unknown process outside the reaction vessel. Here we assume a constant injection rate ρ. (P.3) The rate and shape of the molecule deletion attack is the third input parameter. Unlike in our previous simulations where we performed bulk deletion attacks we now assume a constant attack rate δ at which molecules are randomly picked and destroyed.

The behavior of the system can be characterized by the following three metrics: (M.1) The robustness of the overall system, measured by the probability that the system "survives" a simulation run; (M.2) data yield, expressed as the fraction of injected data molecules x that are not lost, i.e. that are converted to an output molecule y; (M.3) the required CPU power to simulate the system for a given set of input parameters.

Figure 8 qualitatively depicts the influence of input parameters (P.1) − (P.3) to the system metrics (M.1)−(M.3). Each input parameter is assigned an axis of the three-dimensional space, and an arbitrary point in the input parameter space is marked with a dot. The arrows that are leaving the dot indicate an increase of the labled metrics in consequence of changing one of the input parameters. For example, the arrow to the right shows that an increase of the data injection rate ρ requires more CPU power to simulate the system.

Qualitatively, we expect that the robustness of the system and the yield of the data stream can be increased by rising the max. vessel capacity N or by decreasing the rate ρ of injecting input molecules. The robustness is obviously higher if the deletion rate δ is low. More CPU power is needed for higher injection rates or bigger reaction vessels. In the following, we show quantitative measures for these tendencies, obtained by repeated simulation runs.

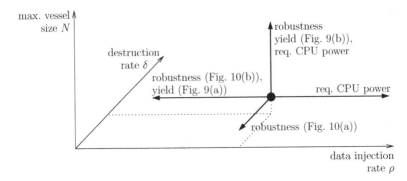

Fig. 8. Qualitative influence of the parameters $(P.1) - (P.3)$ to the metrics $(M.1) - (M.3)$. The figures refered to in parenthesis show the quantitative relations.

Quantitative Considerations. All the results discussed below are obtained by taking the average of 20 simulation runs. For each simulation we inject 10000 input molecules at rate ρ into the reaction vessel of max. capacity N and apply a constant deletion rate δ.

Data Loss. Before focusing on the robustness measure we analyze the yield, i.e. how many of the injected molecules are processed by all operations and finally reach the result vessel. We want to keep the data loss as low as possible, but we expect that even without a deletion attack the excess dilution flow removes some of the molecules belonging to the processed data stream.

The yield is measured by comparing the average rate at which result molecules y are produced (ρ_y) to the rate at which input molecules x are injected (ρ): $yield = \frac{\rho_y}{\rho}$. Figure 9(a) shows the dependency of the data loss on the data injection rate. The concentration of data molecules is very low compared to the concentration of the operation's blueprint and active molecules. Hence data molecules are less frequently selected by the dilution mechanism and therefore, for moderate injection rates ρ, the data loss is surprisingly low despite the excess dilution flow.

When we increase the data injection rate ρ, one instance of the first operation may still be in progress of replicating itself after having processed the previous data molecule. Thus the concentration of its active rules is slightly lower and the reaction rate of the first operation drops. Like this a larger amount of input molecules may accumulate waiting for being processed by the first operation and the probability that the input molecule is removed by the excess dilution flow rises, too. We can also see in Fig. 9(a) that each sequential operation imposes a certain data loss.

We can attenuate the influence of the injection rate ρ to the data yield by increasing the rate at which an operation processes the data stream. This can be done by increasing the concentration of the operation's active molecules, which can be obtained by increasing the max. vessel capacity N. Figure 9(b) shows how an increase of the max. vessel capacity N reduces the overall data loss for a certain data injection rate $\rho = 1$.

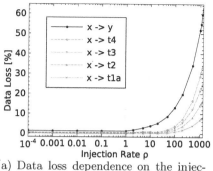

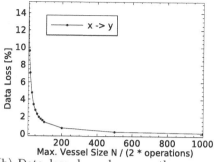

(a) Data loss dependence on the injection rate ρ with fixed $N = 1400$

(b) Data loss dependence on the max. vessel capacity N with fixed $\rho = 1$

Fig. 9. The fraction of data molecules that are lost between the injection of x and an intermediate or final product depends on the injection rate ρ and the max. vessel capacity N. (Here instead of N we use the expected target concentration of blueprint and active molecules for each operation: $\frac{N}{2\omega}$ where $\omega = 7$ is the number of operations.)

Surprisingly, the third input parameter, the rate of deletion attacks, has no effect on the data loss at all as shown in Fig. 10(a) for different data injection rates ρ. However, in this figure we can see that the system becomes unstable when the attack rate rises above a critical level δ_{crit}. Hence we now show how the robustness of the system is influenced by the input parameters.

Robustness. A program shall be called robust if the probability to "survive" a simulation (where 10000 input molecules are injected with the given parameters) is greater than $P_{crit} = 0.9$. In other words, the system is robust if less than 10 % of all simulation runs encounter a starvation situation where no more result molecules y are generated when injecting input data molecules. Starvation might occur when one of the auxiliary molecules necessary for an operation becomes extinct, for example if the number of blueprints drops to zero.

Figure 10(b) depicts the critical attack rate δ_{crit} beyond which the system becomes starved. For low data injection rates ρ the system is able to recover from a deletion attack δ that is about 2ω times higher than ρ, where $\omega = 7$ is the number of operations. When increasing the injection rate, the average number of surviving systems quickly drops to zero. For a lower max. vessel capacity this critical injection rate is much lower.match

Required CPU Power. The CPU power required to execute a program is proportional to the injection rate ρ: The higher the rate of input molecules x the more instructions the algorithm must perform per time interval. In contrast the required CPU power only marginally depend on the vessel capacity N when keeping ρ constant. This is because the replication rate of the quines solely depend on the rate of the data stream. A small effect can be observed because an increase of N causes a cutback of the data loss and thereby more data molecules are able to pass all quines. Therefore when increasing N the number of algorithm cycles spent for each injected molecule only rises slightly.

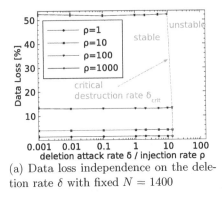
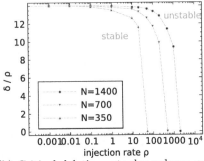

(a) Data loss independence on the deletion rate δ with fixed $N = 1400$

(b) Critical deletion rate dependence on the injection rate ρ with fixed $N = 1400$

Fig. 10. Simulation of a constant random molecule deletion attack with rate δ. Below a critical deletion rate δ_{crit}, the fraction of data molecules that are lost between the injection of x and the final product y does not depend on the deletion rate δ. The critical deletion rate δ_{crit} beyond which the system becomes unstable depends on injection rate.

General Observations. The quantitative analysis of the program showed that the input parameters must be chosen within certain margins in order to obtain a robust program. Obviously, in the probabilistic execution model of an artificial chemistry one never reaches 100 % guarantees for a certain property.

The combination of autocatalytic quines and dilution flow has another stabilizing side effect: If invalid or useless code or data molecules are injected into the reaction vessel, they will eventually be removed by the dilution flow. Therefore the system exhibits some robustness to useless or harmful code. However, this assumption only holds if the injected molecules do not contain self-replicating molecules and if they don't interfere with the operations, as we discuss in following section.

7   From Robustness towards Security

So far we focused on the robustness of Fraglet programs to code and data deletion. In this section we extend the scope and consider further unintentional and malicious attacks to the code and data base.

The exploitation of code has a long history [36]. Over the years many techniques have been developed to exploit code vulnerabilities of common programming mistakes. Thus, these exploits mainly focus on the weaknesses of the code instead of exploiting the theoretical model of the language or the system it runs on. In artificial chemical computing systems the situation is different. Since code is executed via reactions among molecules, it is possible to modify existing molecules such that their computation corresponds to a different functionality. This is based on a characteristic that is deliberately left out of traditional programming languages such as C or Java: Any type of data, input or output data, also represents code and can thus have direct effect on the executed code.

Competing Molecules. If an attacker is allowed to arbitrarily inject molecules into the reaction vessel, he or she can alter the outcome of the computation and destroy the integrity of the program in various ways. For example, the logic of the program presented in Sect. 6 can be changed by injecting competing molecules: The attacker could inject malicious code fraglets ([match t3 mult t4 6]) or data fraglets ([t3 0]), respectively. The first molecule competes against the active fraglet of operation 4 for the input data molecule t3 and masks operation 4 in Fig. 6 to compute $y = 6x^2 + 2x + 1$. The molecule in the latter example directly modifies the result of an operation. However, both attacks are only of a temporary nature because the replicating quines soon reinstall the proper molecules.

Viral Blueprints. A more efficient attack targets the blueprints. An injected viral blueprint [bp1 ...], or an active fraglet [match bp1 bp1...] replaces one of the blueprint instances for the first operation. As long as the concentration of the original blueprint is higher than the viral blueprint the latter will eventually be displaced. But if the attacker injects a large number of modified blueprints the system will activate and replicate the viral blueprint.

Targeted Deletion Attacks. An attack can not only exploit existing code but also disrupt the program. Beyond the analyzed non-selective (blind) deletion attack we can also imagine a targeted deletion attack where an attacker selectively deletes molecules. Even if the attacker has no possibility to directly remove molecules from the reaction vessel he/she can inject active molecules to consume vital passive molecules. For example, the injected active molecule [match bp1 nul] would consume and destroy a blueprint molecule needed to replicate the first operation. Similarly, one could inject molecules ([match t2 nul]) to destroy intermediary results. In turn, all operations that depend on the suppressed data molecule cannot replicate anymore. If either attack is executed with a high number of molecules it drives the system to starvation.

Self-Optimization. Up to now, this section discussed intentional attacks which assumed an active attacker pursuing some specific goal and designing malicious code accordingly. However, the motivation for using artificial chemical computing systems was also based on their potential for self-optimization. In order to better adapt to its environment, a self-optimizing Fraglet program would seek to evolve via genetic programming, by performing random mutation and recombination of its own operations. Such random modifications can lead to code that performs any of the attacks discussed above, and could therefore represent another threat of unintentional attacks. A robustness mechanism as discussed in this paper is an essential building block to make such online self-optimization feasible by filtering out harmful mutants. An explicit fitness-based selection mechanism would also be needed, since those programs that replicate more or faster are not necessarily those that perform the intended tasks.

Concluding Remarks. Our approach of using autocatalytic quines in an artificial chemistry concentrates on a special threat: the non-selective deletion of code and data molecules, which is one of the events that are likely to occur when operating on unreliable devices. We highlighted that there are a vast number of other threats that need to be addressed in future, but we also showed that some of the attacks can already be defended if the number of injected harmful molecules is below a certain limit.

A generic protection mechanism for mobile, self-modifying and self-reproducing code is naturally a big challenge beyond current knowledge. A hypothetical protection mechanism in this context would inherently need to be itself mobile, self-modifying and self-reproducing, such that it could track and respond to such dynamic attack patterns in an effective way. This could lead to an "arms race" of attack and protection waves. The challenge would be to stabilize such a system, in the manner of a biological immune system.

8   Conclusions

In this paper we demonstrated that within Fraglets as an artificial chemistry, it is possible to build self-maintaining program structures that, when exposed to an open but resource constrained environment, exhibit robustness to deletion of their constituent parts. These properties were then brought forward from simple autocatalytic quines to programs consisting of autocatalytic building blocks. Detailed qualitative and quantitative analysis of the program's behaviour showed that the robustness to code and data deletion can be granted within certain input parameter margins.

The presented self-healing mechanism emerges from the combination of a growing population of autocatalytic quines and a random non-selective dilution flow. As a result the code of the program constantly rewrites itself and the dilution flow makes sure that non-replicating instructions vanish.

Acknowledgments

This work has been supported by the European Union and the Swiss National Science Foundation, through FET Project BIONETS and SNP Project Self-Healing Protocols, respectively.

References

1. Baumann, R.C.: Soft errors in advanced semiconductor devices – part 1: the three radiation sources. IEEE Transactions on Device and Materials Reliability 1(1), 17–22 (2001)
2. Anckaert, B., Madou, M., de Bosschere, K.: A model for self-modifying code. In: Camenisch, J.L., Collberg, C.S., Johnson, N.F., Sallee, P. (eds.) IH 2006. LNCS, vol. 4437, pp. 232–248. Springer, Heidelberg (2007)
3. Calude, C.S., Păun, G.: Computing with Cells and Atoms: An Introduction to Quantum, DNA and Membrane Computing. Taylor & Francis, Abington (2001)

4. Dittrich, P.: Chemical Computing. In: Banâtre, J.-P., Fradet, P., Giavitto, J.-L., Michel, O. (eds.) UPP 2004. LNCS, vol. 3566, pp. 19–32. Springer, Heidelberg (2005)

5. Dittrich, P., Ziegler, J., Banzhaf, W.: Artificial Chemistries – A Review. Artificial Life 7(3), 225–275 (2001)

6. Banzhaf, W., Lasarczyk, C.: Genetic Programming of an Algorithmic Chemistry. In: O'Reilly, et al. (eds.) Genetic Programming Theory and Practice II, vol. 8, pp. 175–190. Kluwer/Springer (2004)

7. Deckard, A., Sauro, H.M.: Preliminary Studies on the In Silico Evolution of Biochemical Networks. ChemBioChem 5(10), 1423–1431 (2004)

8. Leier, A., Kuo, P.D., Banzhaf, W., Burrage, K.: Evolving Noisy Oscillatory Dynamics in Genetic Regulatory Networks. In: Collet, P., Tomassini, M., Ebner, M., Gustafson, S., Ekárt, A. (eds.) EuroGP 2006. LNCS, vol. 3905, pp. 290–299. Springer, Heidelberg (2006)

9. Păun, G.: Computing with Membranes. Journal of Computer and System Sciences 61(1), 108–143 (2000)

10. Banâtre, J.P., Fradet, P., Radenac, Y.: A Generalized Higher-Order Chemical Computation Model with Infinite and Hybrid Multisets. In: 1st International Workshop on New Developments in Computational Models (DCM 2005). ENTCS, pp. 5–14. Elsevier, Amsterdam (to appear)

11. Tschudin, C.: Fraglets - a metabolistic execution model for communication protocols. In: Proc. 2nd Annual Symposium on Autonomous Intelligent Networks and Systems (AINS), Menlo Park, USA (2003)

12. Yamamoto, L., Schreckling, D., Meyer, T.: Self-Replicating and Self-Modifying Programs in Fraglets. In: Proc. 2nd International Conference on Bio-Inspired Models of Network, Information, and Computing Systems (BIONETICS 2007), Budapest, Hungary (2007)

13. von Neumann, J.: Theory of Self-Reproducing Automata. University of Illinois Press, Champaign (1966)

14. Sipper, M., Sanchez, E., Mange, D., Tomassini, M., Perez-Uribe, A., Stauffer, A.: A Phylogenetic, Ontogenetic, and Epigenetic View of Bio-Inspired Hardware Systems. IEEE Transactions on Evolutionary Computation 1(1) (1997)

15. Freitas Jr., R.A., Merkle, R.C.: Kinematic Self-Replicating Machines. Landes Bioscience, Georgetown (2004)

16. Sipper, M.: Fifty years of research on self-replication: an overview. Artificial Life 4(3), 237–257 (1998)

17. Langton, C.G.: Self-reproduction in cellular automata. Physica D 10D(1-2), 135–144 (1984)

18. Perrier, J.Y., Sipper, M., Zahnd, J.: Toward a Viable, Self-Reproducing Universal Computer. Physica D 97, 335–352 (1996)

19. Kleene, S.: On notation for ordinal numbers. The Journal of Symbolic Logic 3, 150–155 (1938)

20. Thompson, G.P.: The quine page (1999),
http://www.nyx.net/~gthompso/quine.htm

21. Dittrich, P., di Fenizio, P.S.: Chemical organization theory: towards a theory of constructive dynamical systems. Bulletin of Mathematical Biology 69(4), 1199–1231 (2005)

22. Fontana, W., Buss, L.W.: The Arrival of the Fittest: Toward a Theory of Biological Organization. Bulletin of Mathematical Biology 56, 1–64 (1994)

23. Dittrich, P., Banzhaf, W.: Self-Evolution in a Constructive Binary String System. Artificial Life 4(2), 203–220 (1998)

24. Hutton, T.J.: Evolvable Self-Reproducing Cells in a Two-Dimensional Artificial Chemistry. Artificial Life 13(1), 11–30 (2007)

25. Teuscher, C.: From membranes to systems: self-configuration and self-replication in membrane systems. BioSystems 87(2-3), 101–110 (2007); The Sixth International Workshop on Information Processing in Cells and Tissues (IPCAT 2005), York, UK (2005)

26. Holland, J.: Adaptation in Natural and Artificial Systems, 1st edn. MIT Press, Cambridge (1992)

27. Decraene, J., Mitchell, G.G., McMullin, B., Kelly, C.: The Holland Broadcast Language and the Modeling of Biochemical Networks. In: Ebner, M., O'Neill, M., Ekárt, A., Vanneschi, L., Esparcia-Alcázar, A.I. (eds.) EuroGP 2007. LNCS, vol. 4445, pp. 361–370. Springer, Heidelberg (2007)

28. Tominaga, K., Watanabe, T., Kobayashi, K., Nakamura, M., Kishi, K., Kazuno, M.: Modeling Molecular Computing Systems by an Artificial Chemistry—Its Expressive Power and Application. Artificial Life 13(3), 223–247 (2007)

29. Gillespie, D.T.: Exact Stochastic Simulation of Coupled Chemical Reactions. Journal of Physical Chemistry 81(25), 2340–2361 (1977)

30. Bagley, R.J., Farmer, J.D., Kauffman, S.A., Packard, N.H., Perelson, A.S., Stadnyk, I.M.: Modeling adaptive biological systems. Biosystems 23, 113–138 (1989)

31. Farmer, J.D., Kauffman, S.A., Packard, N.H.: Autocatalytic replication of polymers. Physica D 2(1-3), 50–67 (1986)

32. Kauffman, S.A.: The Origins of Order: Self-Organization and Selection in Evolution. Oxford University Press, Oxford (1993)

33. Mossel, E., Steel, M.: Random biochemical networks: the probability of self-sustaining autocatalysis. Journal of Theoretical Biology 233(3), 327–336 (2005)

34. di Fenizio, P.S., Banzhaf, W.: A less abstract artificial chemistry. In: Bedau, M.A., Mccaskill, J.S., Packard, N.H., Rasmusseen, S. (eds.) Artificial Life VII, Cambridge, Massachusetts 02142, pp. 49–53. MIT Press, Cambridge (2000)

35. Stadler, P.F., Fontana, W., Miller, J.H.: Random catalytic reaction networks. Physica D 63(3-4), 378–392 (1993)

36. Hoglund, G., Mcgraw, G.: Exploiting Software: How to Break Code. Addison-Wesley Professional, Reading (2004)

A Computational Scheme Based on Random Boolean Networks

Elena Dubrova, Maxim Teslenko, and Hannu Tenhunen

Royal Institute of Technology, Electrum 229, 164 46 Kista, Sweden
{dubrova,maximt,hannu}@kth.se

Abstract. For decades, the size of silicon CMOS transistors has decreased steadily while their performance has improved. As the devices approach their physical limits, the need for alternative materials, structures and computational schemes becomes evident. This paper considers a computational scheme based on an abstract model of the gene regulatory network called *Random Boolean Network* (RBN). On one hand, our interest in RBNs is due to their attractive fault-tolerant features. The parameters of an RBN can be tuned so that it exhibits a robust behavior in which minimal changes in network's connections, values of state variables, or associated functions, typically cause no variation in the network's dynamics. On the other hand, a computational scheme based on RBNs seems appealing for emerging technologies in which it is difficult to control the growth direction or precise alignment, e.g. carbon nanotubes.

1 Introduction

A living cell could be considered as a molecular digital computer that configures itself as part of the execution of its code. The core of a cell is the DNA. DNA represents the information for building the basic components of cells as well as encodes the entire process of assembling complex components. By understanding how cells direct the assembly of their molecules, we can find ways to build chips that can self-organize, evolve and adapt to a changing environment.

The *gene regulatory network* is one of the most important signaling networks in living cells [1]. It is composed of the interactions of proteins with the genome. The major discovery related to gene regulatory networks was made in 1961 by French biologists François Jacob and Jacques Monod [2]. They found that a small fraction of the thousands of genes in the DNA molecule acts as tiny "switches". By exposing a cell to a certain hormone, these switches can be turned "on" or "off". The activated genes send chemical signals to other genes which, in turn, get either activated or repressed. The signals propagate along the DNA molecule until the cell settles down into a stable pattern.

Jacob and Monod's discovery showed that the DNA is not just a blueprint for the cell, but rather an automaton which allows for the creation of different types of cells. It answered the long open question of how one fertilized egg cell could differentiate itself into brain cells, lung cells, muscle cells, and other types of cells that form a newborn baby. Each kind of cells corresponds to a different pattern of activated genes in the automaton.

C. Priami et al. (Eds.): Trans. on Comput. Syst. Biol. X, LNBI 5410, pp. 41–58, 2008.
© Springer-Verlag Berlin Heidelberg 2008

In 1969 Stuart Kauffman proposed using *Random Boolean Networks* (RBNs) as an abstract model of gene regulatory networks [3]. Each gene is represented by a vertex in a directed graph. An edge from one vertex to another implies a causal link between the two genes. The "on" state of a vertex corresponds to the gene being expressed. Time is viewed as proceeding in discrete steps. At each step, the new state of a vertex v is a Boolean function of the previous states of the vertices which are the predecessors of v. Kauffman has shown that it is possible to tune the parameters of an RBN so that its statistical features match the characteristics of living cells and organisms [3]. The number of cycles in the RBN's state space, called *attractors*, corresponds to the number of different cell types. Attractor's length corresponds to the cell cycle time. Sensitivity of attractors to different kinds of disturbances, modeled by changing network connections, values of state variables, or associated functions, reflects the stability of the cell to damages, mutations, or virus attacks.

Later RBNs were applied to the problems of cell differentiation [4], immune response [5], evolution [6], and neural networks [7, 8]. They have also attracted the interest of physicists due to their analogy with the disordered systems studied in statistical mechanics, such as the mean field spin glass [9, 10, 11].

In this paper, we investigate how RBNs can be used for computing logic functions. Our interest in RBNs is due, on one hand, to their attractive fault-tolerant features. It is known that parameters of an RBN can be tuned so that the network exhibits a robust behavior, in which minimal changes in network's connections, values of state variables, or associated functions, typically cause no variations in the network's dynamics.

On the other hand, RBNs seem to be an appealing computational scheme for emerging nano-scale technologies in which it is difficult to control the growth direction or achieve precise assembly, e.g. carbon nanotubes. It has been demonstrated that random arrays of carbon nanotubes are much easier to produce compared to the ones with a fixed structure [12]. Random arrays of carbon nanotubes can be deposited at room temperature onto polymeric and many other substrates, which makes them a promising new material for *macroelectronics* applications. Macroelectronics is an important emerging area of technology which aims providing inexpensive electronics on polymeric substrates. Such electronics can be "printed" onto large-area polymeric films by using fabrication techniques similar to text and imaging printing, rather than conventional semiconductor fabrication technology. Applications of macroelectronics include lightweight flexible displays, smart materials or clothing, biological and chemical sensors, tunable frequency-selective surfaces, etc. Conventional semiconductors are not suitable for such applications because they are too expensive and require a crystalline substrate. The effort to develop organic semiconductors has achieved only moderate success so far, mostly because of the low-quality electron transport of organic semiconductors. Random arrays of carbon nanotubes provide a high-quality electron transport and therefore can be a better alternative.

The paper is organized as follows. Section 2 gives a definition of RBNs and summarizes their properties. Section 3 describes how we can use RBNs for computing logic functions and addresses fault-tolerance issues. Section 4 presents a new algorithm for computing attractors in RBNs. Section 5 shows simulation results. Section 6 concludes the paper and discusses open problems.

2 Random Boolean Networks

In this section, we give a brief introduction to Random Boolean Networks. For a more detailed description, the reader is referred to [8].

2.1 Definition of RBNs

A *Random Boolean Network* (RBN) is a synchronous Boolean automaton with n vertices. Each vertex v has k predecessors, assigned independently and uniformly at random from the set of all vertices, and an associated Boolean function $f_v : \{0,1\}^k \rightarrow \{0,1\}$, $k \leq n$. Functions are selected so that they evaluate to values 0 and 1 with given probabilities p and $1 - p$, respectively. Time is viewed as proceeding in discrete steps. At each step, the next value of the state variable x_v associated with the vertex v is a function of the previous values of the state variables x_{u_i} associated with the predecessors of v, u_i, $i \in \{1, 2, \ldots, k\}$:

$$x_v^+ = f_v(x_{u_1}, x_{u_2}, \ldots, x_{u_k}).$$

The state of an RBN is defined by the ordered set of values of the state variables associated with its vertices.

An example of an RBN with $n = 10$ and $k = 2$ is shown in Figure 1. We use "$\cdot$", "$+$" and "$'$" to denote the Boolean operations AND, OR and NOT, respectively.

2.2 Frozen and Chaotic Phases

The parameters k and p determine the dynamics of an RBN. If a vertex controls many other vertices, and the number of controlled vertices grows in time, the RBN is said to be in a *chaotic phase* [13]. Typically such a behavior occurs for large values of $k \sim n$. The dynamics of the network is very sensitive to changes in the values of state variables, associated Boolean function, or network connections.

If a vertex controls only a small number of other vertices and their number remains constant in time, the RBN is said to be in a *frozen phase* [14]. Usually, independently on the initial state, after a few steps, the network reaches a stable state. This behavior usually occurs for small values of k, such as $k = 0$ or 1.

There is a critical line between the frozen and the chaotic phases, on which the number of vertices controlled by a vertex grows in time, but only up to a certain limit [15]. Statistical features of RBNs on the critical line are shown to match the characteristics of real cells and organisms [3, 16]. The minimal disturbances typically create no variations in the network's dynamics. Only some rare perturbations evoke radical changes.

For a given probability p, there is a critical number of inputs k_c below which the network is in the frozen phase and above which the network is in the chaotic phase [9]:

$$k_c = \frac{1}{2p(1-p)}. \tag{1}$$

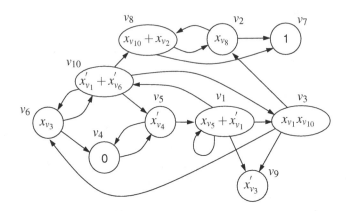

Fig. 1. Example of an RBN with $n = 10$ and $k = 2$. The next state of each vertex v is given by $x_v^+ = f_v(x_{u_1}, x_{u_2})$, where u_1, u_2 are the predecessors of v, and f_v is the Boolean function associated to v. f_v is specified by a Boolean expression written inside the vertex v.

2.3 Attractors

An infinite sequence of consecutive states of a network is called a *trajectory*. A trajectory is uniquely defined by the initial state. Since the number of possible states is finite, all trajectories eventually converge to either a single state, or a cycle of states, called *attractor*. The *basin of attraction* of A is the set of all trajectories converging to the attractor A. The *attractor length* is the number of states in the attractor's cycle.

A number of algorithms for computing attractors in RBNs have been presented. Most of them are based on an explicit representation of the set of states on an RBN and therefore are applicable to networks with up to 32 relevant vertices only [17, 18, 19, 20]. The algorithm presented in [21] uses an implicit representation, namely Binary Decision Diagrams (BDDs) [22], and can handle large RBNs. It finds the set of states of all attractors simultaneously without computing the rest of the states. The individual attractors are then distinguished by simulation. Such an algorithm is a very efficient for RBNs on the critical line, in which the number and length of attractors are of order of the square root of the total number of RBN vertices. The algorithm presented in this paper is intended to complement the algorithm [21] for cases in which the number and/or length of attractors are large and thus distinguishing them by simulation is not efficient.

2.4 Redundant Vertices

It is possible to reduce the state space of an RBN by removing redundant vertices which have no influence on the network's dynamics. A vertex v is considered *redundant* for an RBN G if the reduced network G_R obtained by removing v form G has the same number and length of attractors as G. If a vertex is not redundant, it is called *relevant*.

There are several types of redundant vertices. First, all vertices v whose associated function f_v is constant 0 or constant 1 are redundant. If u is an successor of a redundant

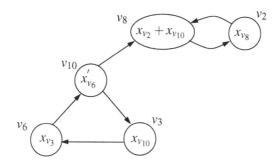

Fig. 2. Reduced network for the RBN in Figure 1

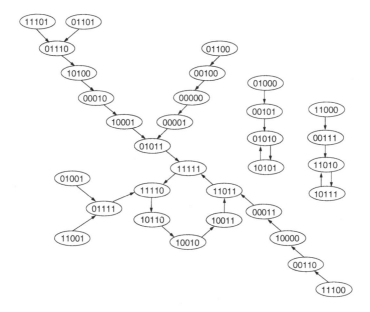

Fig. 3. State transition graph of the RBN in Figure 2. Each state is a 5-tuple $(x_{v_2} x_{v_3} x_{v_6} x_{v_8} x_{v_{10}})$.

vertex v and if after the substitution of the constant value of f_v in f_u the function f_u reduces to a constant, then u is redundant, too.

Second, all vertices v which have no successors are redundant. If all successors of u are redundant, then u is redundant, too.

Third, a vertex can be redundant because its associated function f_v has a constant value due to the correlation of its input variables. For example, if a vertex v with an associated OR (AND) function has predecessors u_1 and u_2 with functions $f_{u_1} = x_w$ and $f_{u_2} = x'_w$, then the value of f_v is always 1 (0). This kind of redundant vertices are hardest to identify.

Exact and approximate bounds on the size of the set of relevant vertices for different values of k and p have been given [23, 14, 15, 13, 24]. In the infinite size limit $n \to \infty$, in the frozen phase, the number of relevant vertices remains finite. In the chaotic phase, the number of relevant vertices is proportional to n. On the critical line, the number of relevant vertices scales as $n^{1/3}$ [17].

The algorithms for computing the set of all redundant vertices, e.g. [18], are too computationally expensive and therefore are feasible for RBNs with up to a thousand vertices only. The *decimation procedure* presented in [20] computes only a subset of redundant vertices, but it is applicable to large networks. In time linear in the size of an RBN it finds redundant vertices evident from the structure of the network (1st and 2nd type). The decimation procedure will not identify the redundant vertices whose associated functions have constant values due to the correlation of their input variables (3rd type).

The reduced network for the RBN in Figure 1 is shown in Figure 2. Its state transition graph is given in Figure 3. Each vertex of the state transition graph represents a 5-tuple $(x_{v_2}x_{v_3}x_{v_6}x_{v_8}x_{v_{10}})$ of values of states on the relevant vertices v_2, v_3, v_6, v_8, v_{10}. There are three attractors: $\{11111, 11110, 10110, 10010, 10011, 11011\}$, $\{01010, 10101\}$ and $\{11010, 10111\}$.

In [25], it has been shown that attractors of an RBN can be computed compositionally from the attractors of the connected components of the reduced network.

3   Computational Scheme Based on RBNs

In this section we discuss how RBNs can be used for computing logic functions. One possibility is to use state variables of relevant vertices of a network to represent variables of the function, and to use attractors to represent the function's values.

To be more specific, suppose that we have an RBN G with r relevant vertices $v_1,\ldots,v_r$ and m attractors $A_1,\ldots,A_m$. The basins of attractions of A_i's partition the Boolean space $\{0,1\}^r$ into m connected components via a dynamic process. Attractors constitute "stable equilibrium" points. We assign a value i, $i \in \{0,1,\ldots,m-1\}$ to the attractor A_i and assume that the set of points of the Boolean space corresponding to the states in the basin of attraction of A_i is mapped to i. Then, G defines a function $f: \{0,1\}^r \to \{0,1,\ldots,m-1\}$ of variables $x_{v_1},\ldots,x_{v_r}$, where the variable x_{v_i} corresponds to the state variable of the relevant vertex v_i. The mapping is unique up to the permutation of m values of f. If $m = 2$, then G represents a Boolean function.

Definition 1. *An RBN with r relevant vertices and m attractors represents a function of type* $f: \{0,1\}^r \to \{0,1,\ldots,m-1\}$ *which is defined as follows:*

$$(a_1,\ldots,a_r) \in B(A_i) \quad \Leftrightarrow \quad f(a_1,\ldots,a_r) = i,$$

$\forall(a_1,\ldots,a_r) \in \{0,1\}^r, \forall i \in \{0,1,\ldots,m-1\}.$

As an example, consider the RBN G shown in Figure 4. The vertices v_4 and v_5 are relevant vertices determining the dynamics of G according to the reduced network in Figure 5(a). The state transition graph of the reduced network is shown in Figure 5(b). There are two attractors, A_0 and A_1. We assign the logic 0 to A_0 and the logic 1 to A_1. The initial states $00, 01$ and 10 terminate in the attractor A_0 (logic 0) and the initial state 11 terminates in the attractor A_1 (logic 1). So, G represents the 2-input Boolean AND.

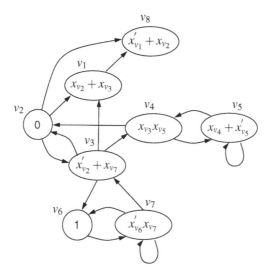

Fig. 4. Example of a network computing the 2-input Boolean AND

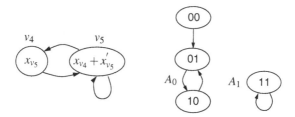

Fig. 5. (a) Reduced network for the RBN in Figure 4. (b) Its state transition graph. Each state is a pair $(x_{v_4} x_{v_5})$. There are two attractors: $A_0 = \{01, 10\}$ and $A_1 = \{11\}$.

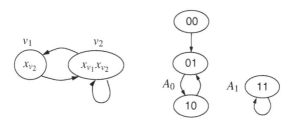

Fig. 6. An alternative reduced network for the 2-input Boolean AND

RBN representation described by the Definition 1 is not unique since we can find many different RBNs representing the same function. For example, the reduced network in Figure 6 has the same state transition graph as the one in Figure 5.

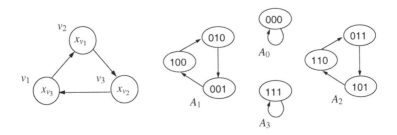

Fig. 7. An (3,1)-RBN for the 3-variable majority function and its state transition graph. The states are ordered as $(x_{v_1}x_{v_2}x_{v_3})$.

One can easily show that an RBN representation with two attractors exists for any n-variable Boolean function, since we can always construct a trivial (n,n)-RBN as follows. Choose any assignment $(a_1,\ldots,a_n) \in \{0,1\}^n$ of variables of f such that $f(a_1,\ldots,a_n) = 0$. Assign $(a_1,\ldots,a_n)$ to be the next state of every state $(b_1,\ldots,b_n)$ of the RBN for which $f(b_1,\ldots,b_n) = 0$.

Similarly, choose any assignment $(c_1,\ldots,c_n) \in \{0,1\}^n$ of variables of f such that $f(c_1,\ldots,c_n) = 1$. Assign $(c_1,\ldots,c_n)$ to be the next state of every state $(d_1,\ldots,d_n)$ of the RBN for which $f(d_1,\ldots,d_n) = 1$.

By construction, the resulting RBN has two single-vertex attractors: $A_0 = (a_1,\ldots,a_n)$ and $A_1 = (c_1,\ldots,c_n)$ and the following associated functions $f_{v_1},\ldots,f_{v_n}$:

- $f_{v_i} = f$ if $a_i = 0$ and $c_i = 1$;
- $f_{v_i} = f'$ if $a_i = 1$ and $c_i = 0$;
- $f_{v_i} = 0$ if $a_i = 0$ and $c_i = 0$;
- $f_{v_i} = 1$ if $a_i = 1$ and $c_i = 1$.

It is desirable to minimize the input degree k of an RBN as much as possible, ideally to $k = 2$, so that a complex functionality is obtained from simpler primitives. However, some functions require $k = n$ for an RBN with two attractors to exist. One example of such function is a 3-variable majority function. Any $(3,k)$-RBN representing it with $k < 3$ has at least 4 attractors (see Figure 7 and Table 1 for an example).

It is possible, however, to find a $(3,3)$-RBN for the majority such that the functions associated to the vertices are simpler than the majority itself. Consider, for instance, the case shown in Table 2 and Figure 8. The vertex v_3 has a 3-input XOR associated to it. Other two vertices are 1- and 2-variable functions.

Rather than increasing the input degree of RBN's vertices, we can work with multiple-valued output functions, or, equivalently, map the same logic value to several attractors. Consider again the RBN in Figure 7. We can assign four different values to the attractors and treat the resulting representation as a function of type $\{0,1\}^3 \rightarrow \{0,1,2,3\}$. Note that this function actually *counts* the number of 1's in the input assignment.

A computational scheme based on RBNs inherits their attractive fault-tolerant features. Many experimental results confirm that RBNs are tolerant to faults, i.e. typically

Table 1. A mapping of RBN's states for the 3-variable majority function resulting in 4 attractors and $k = 1$

x_{v_1}	x_{v_2}	x_{v_3}	$x_{v_1}^+$	$x_{v_2}^+$	$x_{v_3}^+$	f
0	0	0	0	0	0	0
0	0	1	1	0	0	0
0	1	0	0	0	1	0
0	1	1	1	0	1	1
1	0	0	0	1	0	0
1	0	1	1	1	0	1
1	1	0	0	1	1	1
1	1	1	1	1	1	1

the number and length of attractors are not affected by small changes (see [8] for an overview). The following types of fault models are usually used:

- a predecessor of a vertex v is changed to another vertex, i.e. an edge (u,v) is replaced by an edge (w,v), $v,u,w \in V$;
- the value of a state variable is changed to the complemented value;
- Boolean function of a vertex is changed to a different Boolean function.

The stability of the proposed RBN-based computation scheme is due not only to the large percentage of redundancy of RBNs, but also to the non-uniqueness of the RBN representation. As we have shown before, the same function can be implemented by many different RBNs. For instance, the 2-input Boolean AND can be realized in many other ways than the one shown in Figure 4.

Another interesting feature of RBNs is their capability to evolve to a predefined target function. This feature might be of assistance in constructing a good RBN-based representation for a given function. For example, suppose that the following three mutations are applied to the network in Figure 4:

1. the edge (v_4, v_5) is replaced by (v_3, v_5);
2. the edge (v_2, v_3) is replaced by (v_3, v_3);
3. the edge (v_7, v_3) is replaced by (v_5, v_3).

After the removal of redundant vertices from the resulting modified network, we obtain the reduced network shown in Figure 9. Its state space has two attractors, A_0 and A_1. If we assign the logic 0 to A_0 and the logic 1 to A_1, then the initial states 01 and 10 terminate in 1, while 00 and 11 terminate in 0. So, the modified network implements the 2-input Boolean XOR.

4   Computation of Attractors

In order to realize a Boolean function using the method described above, attractors in the states space of an RBN have to be computed. In this section, we show how this can be done by using a technique similar to the fixed point computation in reachability analysis.

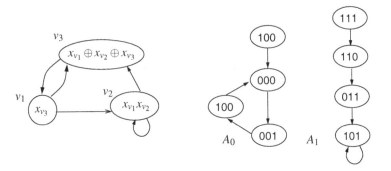

Fig. 8. An (3,3)-RBN for the 3-variable majority function and its state transition graph. The states are ordered as $(x_{v_1}x_{v_2}x_{v_3})$.

Table 2. A mapping of RBN's states for the 3-variable majority function resulting in 2 attractors and $k = 3$

x_{v_1}	x_{v_2}	x_{v_3}	$x_{v_1}^+$	$x_{v_2}^+$	$x_{v_3}^+$	f
0	0	0	0	0	1	0
0	0	1	1	0	0	0
0	1	0	0	0	0	0
0	1	1	1	0	1	1
1	0	0	0	0	0	0
1	0	1	1	0	1	1
1	1	0	0	1	1	1
1	1	1	1	1	0	1

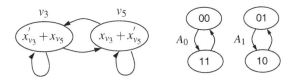

Fig. 9. (a) Reduced network for the RBN in Figure 4, after three mutations described in Section 3 have been applied. (b) Its state transition graph. Each state is a pair $(x_{v_3}x_{v_5})$. There are two attractors: $A_0 = \{00, 11\}$ and $A_1 = \{01, 10\}$.

The main idea can be summarized as follows. Starting from an arbitrary state, forward reachability analysis is applied to find a state in some attractor A. Then, using this state as a final state, backward reachability analysis is performed to find the remaining states in the basin of attraction of A. The process in repeated starting from a state not previously visited until the complete state space is covered.

To be able to compute attractors in large RBNs, it is important to use an efficient representation for their state spaces. The presented algorithm uses reduced ordered Binary Decision Diagrams [22] for representing the set of states of an RBN and the transition relation on this set. Unlike other compressed representations of relations or sets, Binary Decision Diagrams perform the actual operations directly on the compressed representation, without decompressing its first. Before presenting the algorithm, we give a brief introduction to Binary Decision Diagrams, the transition relation, and the traditional reachability analysis.

4.1 Binary Decision Diagrams

A *Binary Decision Diagram (BDD)* is a rooted directed acyclic graph which consists of decision nodes and two terminal nodes called 0- and 1-terminal [22]. Each decision node is labeled by a Boolean variable and has two children called *low* and *high* child. The edge from a node to a low (high) child represents an assignment of the variable to 0 (1). A path from the root node to the 1 (0)-terminal node represents an assignment of variables for which the represented Boolean function evaluates to 1 (0).

A BDD is *ordered* if different variables appear in the same order on all paths from the root to the terminal nodes. A BDD is *reduced* if all isomorphic subgraphs are merged, and any node whose two children are isomorphic is eliminated. The advantage of a reduced ordered BDD is that, for a chosen order of variables, it is unique for the represented function. This property makes reduced ordered BDDs particularly useful in formal verification, since the equivalence of two BDDs can be checked in constant time. Other logical operations, such as conjunction, disjunction, negation, existential quantification, universal quantification, can be performed on BDDs in linear or quadratic time in the size of the graphs [26].

4.2 Transition Relation

A *transition relation* defines the next state values of the vertices in terms of the current state values. We derive the transition relation in the standard way [27], by assigning every vertex v_i of the network a state variable x_{v_i} and making two copies of the set of state variables: $s = (x_{v_1}, x_{v_2}, \ldots, x_{v_r})$, denoting the variables of the current state, and $s^+ = (x_{v_1}^+, x_{v_2}^+, \ldots, x_{v_r}^+)$, denoting the variables of the next state. Using this notation, the characteristic formula for the transition relation of an RBN is given by:

$$T(s, s^+) = \bigwedge_{i=1}^{r} (x_{v_i}^+ \leftrightarrow f_i(x_{v_{i_1}}, x_{v_{i_2}})),$$

where r is the number of relevant vertices, f_i is the Boolean function associated with the vertex v_i and v_{i_1} and v_{i_2} are the predecessors of v_i.

4.3 Forward Reachability

In traditional *forward reachability*, a sequence of formulas $F_i(s)$ representing the set of states that can be reached from a given set of initial states *Init* in i steps is computed as:

$$F_0 = Init,$$
$$F_{i+1}(s^+) = \exists s.(T(s, s^+) \wedge F_i(s)).$$

The sequence generation is terminated when the fixed point is reached for some p:

$$\bigvee_{i=1}^{p} F_i(s) \rightarrow \bigvee_{i=1}^{p-1} F_i(s).$$

In an RBN, *any* state in the state space can be an initial state. We cannot start the reachability analysis from all states as the set of initial states, because then the fixed point is reached immediately. Instead, in our approach, forward reachability is started from a single state selected at random:

$$Init = \bigwedge_{i=1}^{r} (x_{v_i} \leftrightarrow init_i(x_{v_i})),$$

where $init_i(x_{v_i})$ is the initial value of the state variable x_{v_i}, $i \in \{1, \ldots, r\}$.

A sequence of consecutive states from *Init* to an attractor can be quite long (up to 2^r). Thus, it is not efficient to compute $F_i(s)$ for each value of i. To reduce the number of steps needed to reach an attractor, we use the *iterative squaring* technique [28]. Let $T^i(s, s^+)$ denote the transition relation describing the set of next states s^+ that can be reached from any current state s in i steps. For $i = 2$, $T^2(s, s^+)$ is computed as follows:

$$T^2(s, s^+) = \exists s^{++}.(T(s, s^{++}) \wedge T(s^{++}, s^+)). \tag{2}$$

By applying squaring iteratively, we can obtain $T^{2^m}(s, s^+)$ in m steps for any m.

One one hand, it cannot take more than 2^r steps to reach an attractor from any state. One the other hand, "overshooting" is not a problem because, once entered, an attractor is never left. Therefore, for any initial state s, the next state s^+ obtained by the transition defined by $T^{2^r}(s, s^+)$ is a state of an attractor.

We terminate the iterative computation of $T^{2^m}(s, s^+)$ if either m becomes equal to r, or if

$$T^{2^m}(s, s^+) \rightarrow T^{2^{m-1}}(s, s^+),$$

for some $m \in \{1, \ldots, r-1\}$.

Using the resulting transition relation $T^{2^m}(s, s^+)$, $m \in \{1, \ldots, r\}$, we compute the set of states reachable from *Init* in 2^m steps as:

$$F_{2^m}(s^+) = \exists s.(T^{2^m}(s, s^+) \wedge F_0(s)). \tag{3}$$

Note that, with such an approach, we always reach a single state.

4.4   Backward Reachability

The state given by $F_{2^m}(s^+)$ in equation (3) belongs to some attractor A. Next, we perform backward reachability to find the remaining states in the basin of attraction of A.

In traditional *backward reachability*, a sequence of formulas $B_i(s)$ representing the set of states from which a given set of final states *Final* can be reached in i steps is computed as:

$$B_0 = Final,$$
$$B_{i+1}(s) = \exists s^+.(T(s, s^+) \wedge B_i(s^+)).$$

In our case, the set of final states consists of a single state $Final = F_{2^m}(s^+)$ (given by the equation (3)).

The sequence of consecutive states leading to $Final$ can be quite long (up to 2^r). Thus, it is not efficient to compute $B_i(s)$ for each value of i. To reduce the number of steps needed to reach $Final$, we compute a transition relation $T_{0...2^t}(s,s^+)$ which defines the set of all next states s^+ that can be reached from any current state s in $up\ to\ 2^t$ steps. This transition relation is used to obtain the the basin of attraction of A by backward reachability. For $t \in \{0,...,r\}$, $T_{0...2^t}(s,s^+)$ is computed as follows:

$$T_0(s,s^+) = \bigwedge_{i=1}^{r}(x_{v_i}^+ \leftrightarrow x_{v_i}),$$

$$T_{0...1}(s,s^+) = T(s,s^+) \vee T_0(s,s^+),$$

$$T_{0...2^t}(s,s^+) = T_{0...2^{t-1}}^2(s,s^+),$$

where $T_{0...2^{t-1}}^2(s,s^+)$ is computed using the equation (2), and T_0 is the transition relation which assigns the next state of any state to be the state itself.

We terminate the iterative computation of $T_{0...2^t}(s,s^+)$ if either t becomes equal to r, or if

$$T_{0...2^t}(s,s^+) \rightarrow T_{0...2^{t-1}}(s,s^+),$$

for some $t \in \{1,...,r-1\}$.

Using the resulting transition relation, we compute the set of states from which the state $Final$ is reachable in up to 2^t steps as

$$B_{0...2^t}(s) = \exists s^+.(T_{0...2^t}(s,s^+) \wedge B_0(s^+)),$$

where $B_0(s^+) = Final$. The resulting set $B_{0...2^t}(s)$ is the basin of attraction of A.

The whole process is repeated starting from a state not belonging to any previously computed basin of attraction. The algorithm terminates when the complete state space is covered.

The pseudo-code of the algorithm described above is summarized in Figure 10.

Note that we can use the relation $T_{0...2^t}(s,s^+)$ instead of $T_{2^m}(s,s^+)$ when we search for a state of an attractor by forward reachability. In this way, the calculation of $T_{2^m}(s,s^+)$ can be avoided without changing the overall procedure. The only difference is that instead of a single state in an attractor, $F_{2^m}(s^+)$, a set of states is obtained.

4.5 Example

As an example, consider the reduced RBN in Figure 2 and its state transition graph in Figure 3. We have $s = (x_{v_2}, x_{v_3}, x_{v_6}, x_{v_8}, x_{v_{10}})$ and $s^+ = (x_{v_2}^+, x_{v_3}^+, x_{v_6}^+, x_{v_8}^+, x_{v_{10}}^+)$. The transition relation is given by:

$$T(s,s^+) = (x_{v_2}^+ \leftrightarrow x_{v_8}) \wedge (x_{v_3}^+ \leftrightarrow x_{v_{10}}) \wedge (x_{v_6}^+ \leftrightarrow x_{v_3}) \wedge (x_{v_8}^+ \leftrightarrow (x_{v_2}+x_{v_{10}})) \wedge (x_{v_{10}}^+ \leftrightarrow x_{v_6}').$$

After the first iteration of squaring, we get

$$T^2(s,s^+) = \exists s^{++}.((x_{v_2}^{++} \leftrightarrow x_{v_8}) \wedge (x_{v_3}^{++} \leftrightarrow x_{v_{10}}) \wedge (x_{v_6}^{++} \leftrightarrow x_{v_3}) \wedge (x_{v_8}^{++} \leftrightarrow (x_{v_2}+x_{v_{10}}))$$
$$\wedge (x_{v_{10}}^{++} \leftrightarrow x_{v_6}') \wedge (x_{v_2}^+ \leftrightarrow x_{v_8}^{++}) \wedge (x_{v_3}^+ \leftrightarrow x_{v_{10}}^{++}) \wedge (x_{v_6}^+ \leftrightarrow x_{v_3}^{++})$$
$$\wedge (x_{v_8}^+ \leftrightarrow (x_{v_2}^{++}+x_{v_{10}}^{++})) \wedge (x_{v_{10}}^+ \leftrightarrow (x_{v_6}^{++})').$$

algorithm FINDATTRACTORS $(T(s,s^+))$

    number_of_attractors = 0;

    $C(s) = \emptyset$; /* set of states in computed basins of attraction */

    $T^1(s,s^+) = T(s,s^+)$;

    $m = 0$;

    while $m < r$ **do**

        m++;

        $T^{2^m}(s,s^+) = \exists s^{++}.(T^{2^{m-1}}(s,s^{++}) \wedge T^{2^{m-1}}(s^{++},s^+))$;

        if $T^{2^m}(s,s^+) = T^{2^{m-1}}(s,s^+)$ **then**

            break ;

    end while

    $T_0(s,s^+) = \bigwedge_{i=1}^r (x_{v_i}^+ \leftrightarrow x_{v_i})$;

    $T_{0\ldots1}(s,s^+) = T(s,s^+) \vee T_0(s,s^+)$;

    $t = 0$;

    while $t < r$ **do**

        t++;

        $T_{0\ldots2^t}(s,s^+) = T_{0\ldots2^{t-1}}^2(s,s^+)$;

        if $T_{0\ldots2^t}(s,s^+) = T_{0\ldots2^{t-1}}(s,s^+)$;

            break ;

    end while

    while $C(s) \neq U$ **do** /* U is the complete state space */

        Pick up an initial state $F_0(s) \in C'(s)$;

        $F_{2^m}(s^+) = \exists s.(T^{2^m}(s,s^+) \wedge F_0(s))$;

        $B_{0\ldots2^t}(s) = \exists s^+.(T_{0\ldots2^t}(s,s^+) \wedge F_{2^m}(s^+))$;

        $C(s) = C(s) \cup B_{0\ldots2^t}(s)$;

        number_of_attractors++;

    end while

    return (*number_of_attractors*)

end

Fig. 10. Pseudocode of the presented algorithm for computing attractors in RBNs

Similarly, we compute $T^4(s,s^+)$, $T^8(s,s^+)$, and $T^{16}(s,s^+)$. None of $T^{2^m}(s,s^+)$ equals to $T^{2^{m-1}}(s,s^+)$ for $m \in \{1,\ldots,4\}$. So, $T^{32}(s,s^+)$ is computed and the iterative squaring is terminated.

Suppose that the initial state is (01100), i.e.

$$Init = (x_{v_2} \leftrightarrow 0) \wedge (x_{v_3} \leftrightarrow 1) \wedge (x_{v_6} \leftrightarrow 1) \wedge (x_{v_8} \leftrightarrow 0) \wedge (x_{v_{10}} \leftrightarrow 0).$$

Then, by substituting $T^{32}(s,s^+)$ and $F_0(s) = Init$ in (3) and simplifying the result, we get

$$F_{32}(s^+) = (x_{v_2} \leftrightarrow 1) \wedge (x_{v_3} \leftrightarrow 0) \wedge (x_{v_6} \leftrightarrow 0) \wedge (x_{v_8} \leftrightarrow 1) \wedge (x_{v_{10}} \leftrightarrow 0),$$

i.e. the state (10010).

Backward reachability analysis gives us the remaining states in the basin of attraction. By repeating the process starting from, say (01000), we compute the second attractor. By repeating it again starting from, say (11000), we compute the third attractor.

Table 3. Simulation results. Average values for 1000 RBNs with $k = 2$ and $p = 0.5$. "$*$" indicates that the average is computed only for successfully terminated cases.

total number of vertices	average number of vertices	average number of attractors
10	5	2.69
10^2	25	11.3
10^3	93	24.1*
10^4	270	89.7*
10^5	690	-
10^6	1614	-
10^7	3502	-

Since the complete state space is covered by the basins of attraction of the three attractors, the algorithm terminates.

5   Simulation Results

This section shows simulation results for RBNs of sizes from 10 to 10^7 vertices with the parameters $k = 2$ and $p = 0.5$ (Table 3). Column 2 gives the average number of relevant vertices computed using the algorithm from [21]. Column 3 shows the average number of attractors in the resulting networks.

The simulation results show that on RBNs BDDs blow up more frequently than on sequential circuits. Currently, we cannot compute the exact number of attractors in most networks with 10^3 vertices and larger. The number of attractors shown in column 5 for networks with 10^3 and 10^4 vertices is the average value computed for successfully terminated cases only. We did have occasional blow ups for networks with 100 vertices as well. The number of attractors shown in column 5 for networks with 100 vertices is the average value computed for 1000 successfully terminated cases.

6   Conclusion and Open Problems

In this paper, we describe a computational scheme in which states of the relevant vertices of an RBN represent variables of the function, and attractors represent function's values. Such a computational scheme has attractive fault-tolerant features and it seems to be appealing for emerging nanotechnologies in which it is difficult to control the growth direction or achieve precise assembly.

We also present an algorithm for computing attractors in RBN which uses BDDs for representing the set of states of an RBN and the transition relation on this set. However, BDD representation runs out of memory for most networks with 10^3 vertices and larger. We plan to investigate possibilities for implementing the algorithm presented in [21] using Boolean circuits [29], rather than BDDs, and combined approaches [30, 31]. A Boolean circuit is a representation for Boolean formulas in which basic functions are restricted to AND, OR and NOT and all common sub-formulas are shared. Contrary to BDDs, Boolean circuits are not a canonical representation. Therefore, they normally take less memory to be stored, but need more time to be manipulated. We will also try reducing the state space of an RBN by detecting equivalent state variables [32] and by decomposing the transition relation [33].

We also need to find a better technique for partitioning an RBN into components. In our current algorithm, two relevant vertices are assigned to the same component if and only if there is an undirected path between them. Our simulation results show that, for such a partitioning, the size of the largest component of the subgraph induced by the relevant vertices is $\Theta(r)$, where r is the number of relevant vertices in the subgraph. In other words, almost all relevant vertices belong to one connected component. A technique resulting in a more balanced partitioning is needed.

In our future work, we also plan to investigate possibilities for enhancing RBNs as a model. First, input connectivity of gene regulatory networks is much higher than $k = 2$. For example, it is more than 20 in β-globine gene of humans and more than 60 for the platelet-derived growth factor β receptor [8]. We will consider networks with a higher input connectivity k and a smaller probability p, satisfying the equation (1)).

Second, using Boolean functions for describing the rules of regulatory interactions between the genes seems too simplistic. It is known that the level of gene expression depends on the presence of activating or repressing proteins. However, the *absence* of a protein can also influence the gene expression [8]. Using multiple-valued functions instead of Boolean ones for representing the rules of regulations could be a better option.

Third, the number of attractors in RBNs is a function of the number of vertices. However, organisms with a similar number of genes may have different numbers of cell types. For example, humans have 20.000-25.000 genes and more than 250 cell types [34]. The flower Arabidopis has a similar number of genes, 25.498, but only about 40 cell types [35]. We will investigate which factors influence the number of attractors.

Another interesting problem is investigating networks with a different type of connectivity. In RBNs each vertex has an equal probability of being connected to other vertices. Alternatively, in cellular automata [36], each vertex is connected to its immediate neighbors only, and all connections are arranged in a regular lattice. Intermediate cases are possible, e.g. in small-world networks [37] some connections are to distant vertices and some are to neighboring vertices.

References

[1] Alberts, B., Bray, D., Lewis, J., Ra, M., Roberts, K., Watson, J.D.: Molecular Biology of the Cell. Garland Publishing, New York (1994)

[2] Jacob, F., Monod, J.: Genetic regulatory mechanisms in the synthesis of proteins. Journal of Molecular Biology 3, 318–356 (1961)

[3] Kauffman, S.A.: Metabolic stability and epigenesis in randomly constructed nets. Journal of Theoretical Biology 22, 437–467 (1969)

[4] Huang, S., Ingber, D.E.: Shape-dependent control of cell growth, differentiation, and apoptosis: Switching between attractors in cell regulatory networks. Experimental Cell Research 261, 91–103 (2000)

[5] Kauffman, S.A., Weinberger, E.D.: The nk model of rugged fitness landscapes and its application to maturation of the immune response. Journal of Theoretical Biology 141, 211–245 (1989)

[6] Bornholdt, S., Rohlf, T.: Topological evolution of dynamical networks: Global criticality from local dynamics. Physical Review Letters 84, 6114–6117 (2000)

[7] Atlan, H., Fogelman-Soulie, F., Salomon, J., Weisbuch, G.: Random Boolean networks. Cybernetics and System 12, 103–121 (2001)

[8] Aldana, M., Coopersmith, S., Kadanoff, L.P.: Boolean dynamics with random couplings, http://arXiv.org/abs/adap-org/9305001

[9] Derrida, B., Pomeau, Y.: Random networks of automata: a simple annealed approximation. Biophys. Lett. 1, 45 (1986)

[10] Derrida, B., Flyvbjerg, H.: Multivalley structure in Kauffman's model: Analogy with spin glass. J. Phys. A: Math. Gen. 19, L1103 (1986)

[11] Derrida, B., Flyvbjerg, H.: Distribution of local magnetizations in random networks of automata. J. Phys. A: Math. Gen. 20, L1107 (1987)

[12] Snow, E.S., Novak, J.P., Campbell, P.M., Park, D.: Random networks of carbon nanotubes as an electronic material. Applied Physics Letters 81(13), 2145–2147 (2003)

[13] Luque, B., Sole, R.V.: Stable core and chaos control in Random boolean networks. Journal of Physics A: Mathematical and General 31, 1533–1537 (1998)

[14] Flyvbjerg, H., Kjaer, N.J.: Exact solution of Kauffman model with connectivity one. J. Phys. A: Math. Gen. 21, 1695 (1988)

[15] Bastola, U., Parisi, G.: The critical line of Kauffman networks. J. Theor. Biol. 187, 117 (1997)

[16] Kauffman, S.A.: The Origins of Order: Self-Organization and Selection of Evolution. Oxford University Press, Oxford (1993)

[17] Socolar, J.E.S., Kauffman, S.A.: Scaling in ordered and critical random Boolean networks, http://arXiv.org/abs/cond-mat/0212306

[18] Bastola, U., Parisi, G.: The modular structure of Kauffman networks. Phys. D 115, 219 (1998)

[19] Wuensche, A.: The DDlab manual (2000), http://www.cogs.susx.ac.uk/users/andywu/man_contents.html

[20] Bilke, S., Sjunnesson, F.: Stability of the Kauffman model. Physical Review E 65, 016129 (2001)

[21] Dubrova, E., Teslenko, M., Martinelli, A.: Kauffman networks: Analysis and applications. In: Proceedings of the IEEE/ACM International Conference on Computer-Aided Design, pp. 479–484 (November 2005)

[22] Bryant, R.: Graph-based algorithms for Boolean function manipulation. Transactions on Computer-Aided Design of Integrated Circuits and Systems 35, 677–691 (1986)

[23] Flyvbjerg, H.: An order parameter for networks of automata. J. Phys. A: Math. Gen. 21, L955 (1988)

[24] Bastola, U., Parisi, G.: Relevant elements, magnetization and dynamic properties in Kauffman networks: a numerical study. Physica D 115, 203 (1998)

[25] Dubrova, E., Teslenko, M.: Compositional properties of Random Boolean Networks. Physical Review E 71, 056116 (2005)

[26] Hachtel, G.D., Somenzi, F.: Logic Synthesis and Verification Algorithms. Kluwer Academic Publishers, Norwell (2000)

[27] Burch, J., Clarke, E., McMillan, K., Dill, D., Hwang, L.: Symbolic Model Checking: 10^{20} States and Beyond. In: Proceedings of the Fifth Annual IEEE Symposium on Logic in Computer Science, Washington, D.C, pp. 1–33. IEEE Computer Society Press, Los Alamitos (1990)

[28] Burch, J., Clarke, E., Long, D.E., McMillan, K., Dill, D.: Symbolic Model Checking for sequential circuit verification. Transactions on Computer-Aided Design of Integrated Circuits and Systems 13(4), 401–442 (1994)

[29] Bjesse, P.: DAG-aware circuit compression for formal verification. In: Proceedings of the IEEE/ACM International Conference on Computer-Aided Design, pp. 42–49 (November 2004)

[30] Reddy, S.M., Kunz, W., Pradhan, D.K.: Novel verification framework combining structural and OBDD methods in a synthesis environment. In: Proceedings of the 32th ACM/IEEE Design Automation Conference, San Francisco, pp. 414–419 (June 1995)

[31] Williams, P.F., Biere, A., Clarke, E.M., Gupta, A.: Combining decision diagrams and SAT procedures for efficient symbolic model checking. In: Emerson, E.A., Sistla, A.P. (eds.) CAV 2000. LNCS, vol. 1855, pp. 125–138. Springer, Heidelberg (2000)

[32] van Eijk, C.A.J., Jess, J.A.G.: Detection of equivalent state variables in finite state machine verification. In: 1995 ACM/IEEE International Workshop on Logic Synthesis, Tahoe City, CA, pp. 3-35–3-44 (May 1995)

[33] Geist, D., Beer, I.: Efficient model checking by automated ordering of transition relation partitions. In: Dill, D.L. (ed.) CAV 1994. LNCS, vol. 818, pp. 299–310. Springer, Heidelberg (1994)

[34] Liu, A.Y., True, L.D.: Characterization of prostate cell types by cd cell surface molecules. The American Journal of Pathology 160, 37–43 (2002)

[35] Birnbaum, K.D., Shasha, D.E., Wang, J.Y., Jung, J.W., Lambert, G.M., Galbraith, D.W., Benfey, P.N.: A global view of cellular identity in the Arabidopsis root. In: Proceedings of the International Conference on Arabidopsis Research, Berlin, Germany (July 2004)

[36] De Sales, J.A., Martins, M.L., Stariolo, D.A.: Cellular automata model for gene networks. Physical Review E 55, 3262–3270 (1997)

[37] Strogatz, S.H.: Exploring complex networks. Nature 410, 268–276 (2001)

On Channel Capacity and Error Compensation in Molecular Communication

Baris Atakan and Ozgur B. Akan

Next generation Wireless Communications Laboratory
Department of Electrical and Electronics Engineering
Middle East Technical University, 06531, Ankara, Turkey
Tel.: +90(312) 210-4584
{atakan,akan}@eee.metu.edu.tr
http://www.eee.metu.edu.tr/~nwcl/

Abstract. Molecular communication is a novel paradigm that uses molecules as an information carrier to enable nanomachines to communicate with each other. Controlled molecule delivery between two nanomachines is one of the most important challenges which must be addressed to enable the molecular communication. Therefore, it is essential to develop an information theoretical approach to find out communication capacity of the molecular channel. In this paper, we develop an information theoretical approach for capacity of a molecular channel between two nanomachines. Using the principles of mass action kinetics, we first introduce a molecule delivery model for the molecular communication between two nanomachines called as Transmitter Nanomachine (TN) and Receiver Nanomachine (RN). Then, we derive a closed form expression for capacity of the channel between TN and RN. Furthermore, we propose an adaptive Molecular Error Compensation (MEC) scheme for the molecular communication between TN and RN. MEC allows TN to select an appropriate molecular bit transmission probability to maximize molecular communication capacity with respect to environmental factors such as temperature and distance between nanomachines. Numerical analysis show that selecting appropriate molecular communication parameters such as concentration of emitted molecules, duration of molecule emission, and molecular bit transmission probability it can be possible to achieve high molecular communication capacity for the molecular communication channel between two nanomachines. Moreover, the numerical analysis reveals that MEC provides more than % 100 capacity improvement in the molecular communication selecting the most appropriate molecular transmission probability.

Keywords: Molecular communication, nanomachines, molecular bit, information theory, channel capacity, error compensation.

1 Introduction

Molecular Communication is a new interdisciplinary research area including the nanotechnology, biotechnology, and communication technology [1]. In nature, molecular communication is one of the most important biological functions in

C. Priami et al. (Eds.): Trans. on Comput. Syst. Biol. X, LNBI 5410, pp. 59–80, 2008.
© Springer-Verlag Berlin Heidelberg 2008

living organisms to enable biological phenomena to communicate with each other. For example, in an insect colony, insects communicate with each other by means of pheromone molecules. When an insect emits the pheromone molecules, some of them bind the receptors of some insects in the colony and these insects convert the bound pheromone molecules to biologically meaningful information. This enables the insects in the colony to communicate with each other. Similar to insects, almost all of the biological systems in nature perform intra-cellular communication through vesicle transport, inter-cellular communication through neurotransmitters, and inter-organ communication through hormones [1].

As in nature, molecular communication is also indispensable to enable nano-scale machines to communicate with each other. Nanotechnology is one of the most important promising technology which enables nano-scale machines called as nanomachines. Nanomachines are molecular scale objects that are capable of performing simple tasks such as actuation and sensing [1]. Nanomachines are categorized into two types. While one type mimics the existing machines, other type mimics nature made nanomachines such as molecular motors and receptors [2]. In the biological systems, communication among the cells forming the biological system is essential to enable the cells to effectively accomplish their tasks. For example, in natural immune system, the white blood cells called as B-cells and T-cells communicate with each other to eliminate the pathogen entering the body. Similar to biological systems, communication among nanomachines is essential for effective sensing and action.

Due to size and capabilities of nanomachines, the traditional wireless communication with electromagnetic waves cannot be possible to communicate nanomachines that constitute of just several hundreds of atoms or molecules [1]. Instead, the molecular communication is a viable communication paradigm, which enables nanomachines to communicate with each other using molecules as information carrier [1]. Therefore, a molecular channel is envisioned as a communication channel for the molecular communication between two nanomachines. For this channel, it is essential to find out molecular delivery capacity between two nanomachines to understand how to enable molecular communication with high molecule delivery capacity. The molecule delivery capacity may be affected by some parameters specific to the nanomachines and physical properties of the environment such as diffusion coefficient and temperature. Therefore, it is imperative to find out capacity of the molecular channel and to understand how it varies with the properties of the nanomachines and environment.

There exist several research efforts about the molecular communication in the literature. In [1], research challenges in molecular communication is manifested. In [3], the concept of molecular communication is introduced and first attempt for design of molecular communication system is performed. In [4], a molecular motor communication system for molecular communication is introduced. In [5], a molecular communication system which will enable future health care applications is investigated. In [6], based on intercellular calcium signaling networks, the design of a molecular communication system is introduced. In [7], an autonomous molecular propagation system is proposed to transport information molecules using DNA hybridization and biomolecular linear motors. The existing studies on the molecular communication include feasibility of the molecular communication and design schemes for molecular communication system. However, none of these studies investigate the capacity of a molecular channel to understand possible conditions in which the molecular communication can be feasible and high molecular communication capacity can be achieved.

In this paper, we introduce an information theoretical approach for molecular communication and propose a closed form expression for molecular communication capacity between two nanomachines and propose an adaptive error compensation technique for molecular communication by significantly extending our preliminary work in [8]. Using the principles of mass action kinetics, we first model the molecular delivery between two nanomachines called Transmitter Nanomachine (TN) and Receiver Nanomachine (RN). Then, based on the molecular delivery model, we derive the closed form expression for capacity of the channel between TN and RN. In this paper, we also propose an adaptive Molecular Error Compensation (MEC) scheme for the molecular communication between TN and RN. We first define an interval for selection of the most appropriate molecular bit transmission probability providing higher molecular communication capacity with minimum error. Then, using this interval, we introduce a selection strategy to enable TN to select the most appropriate molecular bit transmission probability with respect to some environmental factors such as temperature, binding rate, and distance between nanomachines. MEC allows TN and RN to collaboratively select the most appropriate molecular bit transmission probability providing high molecular communication capacity. Finally, using the capacity expression and the error compensation scheme, we investigate how the conditions such as temperature of environment, concentration of emitted molecules, distance between nanomachines and duration of molecule emission affect the molecular communication capacity and molecular bit transmission probability that provides higher molecular communication capacity. We further discuss under which conditions the molecular communication can be feasible with high capacity.

The remainder of this paper is organized as follows. In Section 2, we introduce a molecular communication model. In Section 3, we introduce a molecule delivery approach for the molecular communication between two nanomachines. In Section 4, based on the molecule delivery scheme we introduce an information theoretical approach for the molecular communication between two nanomachines. In Section 5, we propose an adaptive error compensation scheme. In Section 6, we provide the numerical results and we give concluding results in Section 7.

2   Molecular Communication Model

In nature, molecular communication among biological entities is based on the ligand-receptor binding mechanism. According to ligand-receptor binding mechanism, ligand molecules are emitted by one biological phenomenon then, the emitted ligand molecules diffuse in the environment and bind to the receptors of

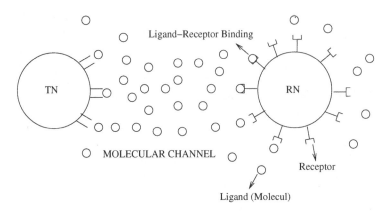

Fig. 1. Molecular Communication Model

another biological phenomenon. This binding enables the biological phenomenon to receive the bound molecules by means of the diffusion on cell membrane. The received ligand molecules allow the biological phenomenon to understand the biological information. For example, in biological endocrine system, gland cells emit hormones to inter-cellular environment then, hormone molecules diffuse and are received by corresponding cells. According to the type of emitted hormone, the corresponding cells convert the hormone molecule to biologically meaningful information. This natural mechanism provides the molecular communication for almost all biological phenomena.

In this paper, we adopt this natural ligand-receptor binding mechanism to enable the molecular communication between nanomachines analogous to the biological mechanisms and called Transmitter Nanomachine (TN) and Receiver Nanomachine (RN) as shown in Fig. 1. In the literature, artificial ligand-receptor binding schemes have been previously introduced [9], [10]. In this paper, we assume an artificial ligand-receptor binding model developed in [9]. We assume that TN is a nano-scale machine or a biological entity and it can emit one kind of molecule called A. We also assume that TN emits molecules A with a time-varying concentration of $L(t)$ according to the following emission pattern [10] which is similar to alternating square pulse, i.e.,

$$L(t) = \begin{cases} L_{ex}, & for \quad jt_H \leq t \leq (j+1)t_H \\ 0, & otherwise \end{cases} \tag{1}$$

where $j = (0, 1, ...)$, t_H is the duration of a pulse and L_{ex} is concentration of molecules A emitted by TN. Furthermore, we assume that RN is a nano-scale machine and it has N receptors called R on its surface. The receptors enable RN to receive the molecules which bind their surface.

In traditional digital communication, information sequences are transmitted via two bits, logic 1 and 0. If a transmitter detects a voltage level which is greater than a prescribed voltage level in the channel, it decides that transmitter transmitted logic 1. If the voltage level in the channel is less than the prescribed level, the receiver decides that the transmitter transmitted logic 0. Using this traditional idea, we propose a similar molecular communication scheme. According to this scheme, during time interval t_H, TN either emits molecules A corresponding to logic 1 in digital communication or it transmits no molecule corresponding to logic 0 in digital communication. If a TN intends to transmit molecules A, we assume that during the time interval t_H, it emits molecules A to its surrounding environment with a specific concentration L_{ex}. Similar to logic 1 and logic 0 in traditional digital communication, we denote the case that TN transmits molecules A with A and we denote the case that TN transmits no molecule with 0. Hence, for the molecular communication model, we have two molecular communication bits called A and 0.

At RN side, these bits are inferred via concentration of molecules A such that if an RN receives a concentration of molecules A greater than a prescribed concentration S ($\mu mol/liter$), the RN decides that the TN transmitted molecular bit A during the time interval t_H. Conversely, if the RN receives molecules A with a concentration less than S, the RN decides that the TN transmitted molecular bit 0.

In traditional digital communication, noise level in the channel leads to channel errors such that when a transmitter intends to transmit logic 0, the receiver may detect logic 1, or for logic 1, the receiver may detect logic 0 due to the noise in the channel. Similarly, in the molecular communication, it may be possible to observe erroneous molecular communication bits at the RN side. During the molecular communication, the molecules A are emitted by TN and the emitted molecules continuously diffuse to surrounding environment including the RN such that molecules A always exist and diffuse in the environment. Therefore, due to the emitted molecules A which diffuse in the surrounding environment, it is possible for RN to receive molecular bit A although TN transmits molecular bit 0. Furthermore, due to delay in diffusion of molecules A to RN it is also possible for RN to receive molecular bit 0 although TN transmits molecular bit A. Moreover, erroneous molecular bits can arise due to some additional factors which affect the molecular diffusion between TN and RN, such as temperature of the environment, concentration of emitted molecules A, distance between TN and RN, duration of molecule emission, binding and release rates, and number of receptors on RN.

Consequently, similar to traditional digital communication channel, the molecular communication channel between TN and RN has a molecule delivery capacity which is defined as maximum number of non-erroneous molecular bits which can be delivered within a specific time duration.

Next, we introduce a molecule delivery model for the molecular communication between TN and RN according to the molecular communication approach introduced here.

3   Molecule Delivery

For the molecular communication between TN and RN, it is important to understand how molecules A can be delivered to RN by means of the binding between molecules A and receptors R on the RN. In this section, assuming the ligand-receptor binding model in [9], we introduce a molecule delivery model for the molecular communication between TN and RN.

According to the ligand-receptor binding reaction kinetics, when molecules A, emitted by TN, encounter with receptors R on RN, molecules A bind to the receptors R. These bound molecules A and receptors R constitute complexes C (bound receptors) according to the following chemical reaction,

$$A + R \xrightarrow{k_1} C \tag{2}$$

where k_1 ($\mu mol/liter/sec.$) is the rate of binding reaction. Similar to the binding reaction, it is possible to release molecules A from receptors R according to the following chemical reaction,

$$A + R \xleftarrow{k_{-1}} C \tag{3}$$

where k_{-1} ($1/sec.$) is rate of release reaction.

As given in (1), TN emits molecules A via a square pulse with amplitude L_{ex} during t_H ($sec.$). In this duration, concentration of bound receptors $C(t)$ ($\mu mol/liter$) can be given [9] as follows

$$C(t) = C_\infty (1 - e^{-t(k_{-1}+k_1 L_{ex})}) \tag{4}$$

where k_1 and k_{-1} are the binding and release rates, respectively, L_{ex} ($\mu mol/liter$) is concentration of molecules A which is emitted by TN. C_∞ is steady-state level of bound receptors and can be given [9] as follows

$$C_\infty = \frac{k_1 L_{ex} N}{k_{-1} + k_1 L_{ex}} \tag{5}$$

where N ($\mu mol/liter$) is the concentration of receptors (R) on RN.

During the pulse duration t_H, $C(t)$ rises exponentially according to (4). At time t_0 when the pulse duration ends, $C(t)$ starts to decay [9] according to

$$C(t) = C_{t_0} e^{[-k_{-1}(t-t_0)]} \; for \; t > t_0 \tag{6}$$

The rates of molecule/receptor interaction, k_1 and k_{-1}, may depend on molecular diffusion from TN to RN. More specifically, while the binding rate k_1 heavily depends on the molecular diffusion parameters from TN to RN such as diffusion coefficient, temperature of environment, distance between TN and RN [11], the release rate k_{-1} depends on some environmental factors such as interaction range and temperature [12]. Here, we only assume that binding rate (k_1)[1] is inversely proportional with distance (α) between TN and RN such that $k_1 \propto 1/\alpha$ and it is directly proportional with temperature of environment (T) such that $k_1 \propto 2T$. For the release rate k_{-1}, we use the model given in [12] as follows

$$k_{-1} = k_{-1}^0 e^{\alpha f/k_B T} \tag{7}$$

---

[1] Here, we do not predict k_1 according to the diffusion parameters of the environment. In fact, binding rate k_1 can be captured with analytical expressions [14]. However, this is out of scope of this paper.

where k^0_{-1} is the zero-force release rate, α is the distance between TN and RN, k_B and T are the Boltzmann constant and absolute temperature, respectively. f is the applied force per bound. f is related with the energy of the emitted molecules, the distance between TN and RN, and the environmental factors [13]. Here, we consider f as positive constant throughout this paper. k^0_{-1} can be predicted by fitting the experimental measurements [12] and it is related with the capability of molecule capturing of RN receptors. Therefore, we assume that k^0_{-1} is a variable which depends only on the properties of RN receptors.

Based on the models introduced in Section 2 and 3, we next develop an information theoretical approach for the capacity of the molecular channel between TN and RN. According to total concentration of complex molecules $(C(t))$ expressed in (4), (5) and (6), we derive probability of erroneous molecular bits which cannot be successfully delivered to RN. Then, we model the molecular communication channel similar to binary symmetric channel and we derive a capacity expression of the molecular channel between TN and RN.

4   An Information Theoretical Approach for Molecular Communication

As introduced in Section 2, for the molecular communication between TN and RN, two molecular bits are available. Every time when TN transmits a molecular bit, concentration of delivered molecules determines the success of the transmission. If TN transmits molecular bit A, at least S number of molecules[2] A must be delivered to RN within time interval t_H for a successful delivery of a molecular bit A. If TN transmits molecular bit 0, number of molecules A delivered within t_H must be less than S for a successful delivery of molecular bit 0. Therefore, it is imperative to find the number of delivered molecules in each transmission interval t_H to determine the success of the molecular bit transmission from TN to RN.

Here, using (4), (5), (6) and (7), the closed form expressions for expected value of number of delivered molecules A during t_H, i.e., N_A, can be given by

$$N_A = \int_0^{t_H} C(t)dt \tag{8}$$

$$N_A = \int_0^{t_H} \frac{k_1 L_{ex} N}{k_{-1} + k_1 L_{ex}} (1 - e^{-t(k_{-1}+k_1 L_{ex})})dt \tag{9}$$

Since the molecular diffusion continues after every t_H interval, the previous molecular bits can be received in the current interval by RN. Therefore, the number of delivered molecules A in a given interval also depends on molecular bits transmitted in the previous intervals. Here, we assume that the last molecular

---

[2] Since concentration of molecules $(\mu mol/liter)$ can be converted to number of molecules by multiplying Avagadro constant (6.02×10^{23}), we interchangeably use the number of molecules for the concentration of molecules.

bit only affects the current molecular transmission since the number of delivered molecules exponentially decay after t_H seconds according to (6). If we assume that TN emits molecules A with probability P_A in each time interval t_H and it emits molecular bit 0 with probability $(1-P_A)$. Hence, the effect of the last emitted molecular bit on the current molecular bit transmission can be considered as expected number of delivered molecules coming from the previous interval, i.e., N_p. Thus, using (6) and (9), N_p can be given as follows

$$N_p = \int_0^{t_H} P_A N_A e^{(-k_{-1}t)} dt \tag{10}$$

$$N_p = \int_0^{t_H} \left(P_A \int_0^{t_H} \frac{k_1 L_{ex} N}{k_{-1} + k_1 L_{ex}} (1 - e^{-t(k_{-1}+k_1 L_{ex})}) dt \right) e^{(-k_{-1}t)} dt \tag{11}$$

Combining (9) and (11), for the case that TN emits A during t_H, expected value of total number of delivered molecules A, i.e., $E[N_{TA}]$, can be given as follows

$$E[N_{TA}] = N_A + N_p \tag{12}$$

At the RN side, if RN receives S number of molecules A, it infers that TN emitted the molecular bit A during t_H. Thus, using the well-known Markov inequality, we obtain a maximum bound for the probability p_1 that TN achieves to deliver molecular bit A as follows

$$p_1(N_{TA} \geq S) \leq \frac{E[N_{TA}]}{S} \tag{13}$$

Hence, TN achieves to deliver molecular bit A with maximum probability $p_1 = \frac{E[N_{TA}]}{S}$ and RN receives molecular bit 0 instead of the molecular bit A such that TN does not succeed to deliver A with probability $(1 - p_1)$.

For the transmission of molecular bit 0 during t_H, the number of delivered molecules A only depends on lastly emitted molecular bit since TN transmits no molecules during the transmission of molecular bit 0. Therefore, following (11), the expected value of total number of delivered molecules A within t_H for the transmission of molecular bit 0, i.e., $E[N_{T0}]$, is given by

$$E[N_{T0}] = N_p \tag{14}$$

For the successful delivery of a molecular bit 0, TN must deliver a number of molecules A that is less than S and $(N_{T0} \leq S)$ to RN. Using the Markov inequality, the maximum bound for probability p_2 that TN achieves to deliver molecular bit 0 is given by

$$p_2(N_{T0} \leq S) \leq \frac{S}{E[N_{T0}]} \tag{15}$$

Hence, for the transmission of molecular bit 0, TN achieves to deliver molecular bit 0 with maximum probability $p_2 = \frac{S}{E[N_{T0}]}$ and it does not achieve to deliver molecular bit 0, instead, it incorrectly delivers molecular bit A with probability $(1 - p_2)$. Here, it is critical to select an appropriate S to maximize p_1 and p_2. As seen in (13), for $E[N_{TA}] \geq S$, $p_1 \geq 1$ is obtained although it is not possible for p_1 to take a value greater than 1. This implies that $p_1 \approx 1$ can be obtained by appropriately selecting S. For example, if we assume that N_{TA} is a random variable with the normal distribution $N(E[N_{TA}], \sigma_{TA}2)$, $p_1 \approx 1$ can be obtained by selecting S as $0 < S < E[N_{TA}] - 3\sigma_{TA}$. This is because in any normal distribution, % 99.7 of the observations fall within 3 standard deviations of the mean. Similarly, if we assume that N_{T0} has the normal distribution $N(E[N_{T0}], \sigma_{T0}2)$, $p_2 \approx 1$ can be succeed selecting S as $S > E[N_{T0}] + 3\sigma_{T0}$. Consequently, p_1 and p_2 can be maximized by selecting S from the interval $0 < E[N_{T0}] + 3\sigma_{T0} < S < E[N_{TA}] - 3\sigma_{TA}$

According to the transmission probabilities p_1 and p_2, we can model a channel similar to the symmetric channel [15]. If we consider that TN emits molecular bit X and RN receives molecular bit Y, then the transition matrix of the molecular channel can be given as follows

$$P(Y/X) = \begin{pmatrix} p_1 P_A & (1 - p_2)(1 - P_A) \\ (1 - p_1)P_A & p_2(1 - P_A) \end{pmatrix}$$

Based on the transition matrix $P(Y/X)$, we can give the mutual information $I(X; Y)$ between X and Y which states the number of distinguishable molecular bits, i.e., M as follows

$$M = \left(H\Big(p_1 P_A + (1 - p_2)(1 - P_A), (1 - p_1)P_A + p_2(1 - P_A)\Big) \right) - \tag{16}$$
$$- \Big(P_A H(p_1, 1 - p_1) + (1 - P_A)H(p_2, 1 - p_2) \Big)$$

$$M = -\left[P_A \tfrac{E[N_{TA}]}{S} + \left(1 - P_A\right)\left(1 - \tfrac{S}{E[N_{T0}]}\right) \right] log \left[P_A \tfrac{E[N_{TA}]}{S} + \left(1 - P_A\right)\left(1 - \tfrac{S}{E[N_{T0}]}\right) \right] - \tag{17}$$
$$- \left[P_A \left(1 - \tfrac{E[N_{TA}]}{S}\right) + \left(1 - P_A\right)\tfrac{S}{E[N_{T0}]} \right] log \left[P_A \left(1 - \tfrac{E[N_{TA}]}{S}\right) + \left(1 - P_A\right)\tfrac{S}{E[N_{T0}]} \right] -$$
$$- P_A \left[\tfrac{E[N_{TA}]}{S} log \left(\tfrac{E[N_{TA}]}{S}\right) - \left(1 - \tfrac{E[N_{TA}]}{S}\right) log \left(1 - \tfrac{E[N_{TA}]}{S}\right) \right] -$$
$$- (1 - P_A) \left[\tfrac{S}{E[N_{T0}]} log \left(\tfrac{S}{E[N_{T0}]}\right) - \left(1 - \tfrac{S}{E[N_{T0}]}\right) log \left(1 - \tfrac{S}{E[N_{T0}]}\right) \right]$$

where $H(.)$ denotes the entropy. Using (17), the capacity of the molecular channel between TN and RN i.e., C_M, can be expressed as

$$C_M = max(M) \tag{18}$$

Traditionally, in a communication channel it is necessary to design codes that enable minimum error rate and maximum capacity. Similarly, in a molecular communication channel, it is necessary to find a molecular bit transmission probability that can maximize molecular communication capacity. Next, we introduce an adaptive error compensation scheme in the molecular communication, which enable TN to select most appropriate molecular bit transmission probability that can maximize molecular communication capacity.

5 Adaptive Error Compensation in the Molecular Communication

In the traditional digital communication, erroneous bits are frequently observed due to noise in the channel. To compensate bit errors, various kinds of channel coding schemes have been proposed in the literature. Generally, the aim of these techniques is to reduce bit error probability of communication between transmitter and receiver. For this aim, transmitter organizes transmitting communication bits to generate fixed-length codewords such that these codewords enable the receiver to detect and correct the erroneous communication bits. However, detection and correction of erroneous communication bits necessitate efficient processors, algorithms, and circuits with high computational power at the receiver side.

As in traditional digital communication, two molecular communication bits are available for the communication between TN and RN. However, existing channel coding techniques are not appropriate for the molecular communication since they necessitate high computational power, which may not be realizable for nanomachines with limited computational and storage capabilities. Therefore, the molecular communication needs proactive error compensation schemes, which do not necessitate any computational processing to compensate possible errors on the molecular channel. These error compensation schemes should proactively prevent the possible errors on the molecular communication channel by adapting some molecular communication parameters according to changing environmental factors such as temperature, binding rate, and distance between nanomachines.

In the molecular communication, we assume that some communication parameters such as concentration of emitted molecules A (L_{ex}), duration of emission pulse (t_H), and concentration of receptors on RN (N) are specific to the TN and RN and related with the design issues of the nanomachines. Therefore, they cannot be changed by neither nanomachines nor environmental factors. However, other parameters such as temperature of the environment (T), applied force per bound (f), distance between TN and RN (α), binding rate (k_1), release rate (k_{-1}), and zero-force release rate (k^0_{-1}) only depend on some environmental factors such as diffusion coefficients of the environment and deployment strategies of the nanomachines. Here, we assume that the probability of molecular bit A emission (P_A) can only be changed by TN. Therefore, a proactive error compensation scheme exploits regulation of P_A such that the regulation can proactively compensate the possible channel errors, which stem from some environmental factors and some properties of the nanomachines. For example, in an environment generating high binding rate (k_1), transmission of molecular bit 0 can be erroneous since high amount of molecules A can be delivered to RN during the transmission of molecular bit 0. However, selection of the most appropriate P_A decreasing the number of delivered molecules can enable TN to compensate such kind of errors.

Theoretically, it is possible to optimize (17) to find a molecular transmission probability (P_A), which minimizes the errors on molecular communication channel and maximizes molecular communication capacity. This can enable TN to encode transmitted molecular bit such that TN can minimize the errors. However, this kind of optimization process is computationally impossible for TN. In stead of some optimization process with high computational burden, it is possible to find some simple methods that enable TN to decide which P_A is the most appropriate in which environmental conditions.

According to the molecular communication model introduced in Section 2, to successfully deliver molecular bit A TN must deliver at least S number of molecules A to RN. Therefore, the condition

$$E[N_{TA}] = N_A + N_p > S \tag{19}$$

must hold for successful delivery of molecular bit A. Substituting N_p given in (10), we rewrite (19) as

$$N_A + P_A N_A \int_0^{t_H} e^{(-k_{-1}t)} dt > S \tag{20}$$

Using (20), a lower bound for P_A, i.e., LB, can be given as

$$P_A > \frac{S - N_A}{N_A \int_0^{t_H} e^{(-k_{-1}t)} dt} = LB \tag{21}$$

where $N_A \int_0^{t_H} e^{(-k_{-1}t)} dt$ states concentration of molecules A that are received by RN within an exponential decaying phase after TN transmits molecular bit A as introduced in (11). We denote $N_A \int_0^{t_H} e^{(-k_{-1}t)} dt$ with N_{ex} and rewrite (21) as

$$LB = \frac{S - N_A}{N_{ex}} \tag{22}$$

To successfully deliver molecular bit 0, TN must deliver a number of molecules A less than S to RN. Therefore, the condition that must be met for successful delivery of molecular bit 0 is expressed as

$$E[N_{T0}] = N_p \leq S \tag{23}$$

where $E[N_{T0}]$ and N_p denote the expected number of delivered molecules coming from the previous interval as introduced in (10). Using (10), (23) can be rewritten as

$$P_A N_A \int_0^{t_H} e^{(-k_{-1}t)} dt \leq S \tag{24}$$

Similar to lower bound given in (22), using (24), an upper bound for P_A, i.e., UB, can be given as follows

$$P_A \leq \frac{S}{N_A \int_0^{t_H} e^{(-k_{-1}t)} dt} = UB \tag{25}$$

$$UB = \frac{S}{N_{ex}} \tag{26}$$

Combining the lower and upper bounds given in (22) and (26), respectively, an interval for selection of the most appropriate P_A that minimizes the channel errors in the molecular communication can be stated as

$$\frac{S - N_A}{N_{ex}} < P_A \leq \frac{S}{N_{ex}} \tag{27}$$

$$LB < P_A \leq UB \tag{28}$$

N_{ex} includes an integral operation that is impossible for TN to practically compute due to its very limited computational power. Since N_{ex} states concentration of molecules A that are received by RN within an exponential decaying phase after TN transmits molecular bit A, RN can obtain N_{ex} by computing concentration of molecules A within an exponential decaying phase after TN transmits a molecular bit A. Here, we assume that similar to the molecular communication from TN to RN, molecular communication from RN to TN can be possible[3]. We also assume that RN computes the concentration within an exponential decaying phase after TN transmits a molecular bit A and it communicates this concentration to RN before initiating the molecular communication. Thus, to determine an appropriate P_A providing satisfactory molecular communication capacity, TN evaluates the lower and upper bounds given in (27). However, TN needs a selection strategy to select the most appropriate molecular bit transmission probability (P_A) from the interval given in (28).

5.1   A Selection Strategy for Molecular Bit Transmission Probability

For the molecular communication, it is critical to select the most appropriate molecular bit transmission probability (P_A) providing high molecular communication capacity. Using the interval given in (27), we investigate how the variation of P_A affects the error rate in the molecular communication such that we derive a selection strategy for P_A, which allows TN to minimize error rate and to maximize molecular communication capacity.

While P_A increases, the number of delivered molecules increases. Therefore, higher P_A decreases the errors in transmission of molecular bit A. Conversely, while P_A increases, errors increase in transmission of molecular bit 0. However, P_A less than UB enables non-erroneous molecular bit 0 transmission as introduced in derivation of UB. Therefore, P_A should be selected as high as possible for non-erroneous molecular bit A transmission while it should be selected less than UB for non-erroneous molecular bit 0. Hence, P_A should be selected as a value that is the closest to UB ($P_A \cong UB$). This selection strategy can minimize

---

[3] Here, we also note that TN and RN have the same molecule delivery and reception capability. However, we do not assume full duplex molecular communication between TN and RN such that TN and RN cannot simultaneously deliver or receive molecular bits. Hence, we assume a half duplex molecular communication between TN and RN.

error rate and maximize molecular communication capacity. According to this P_A selection strategy, UB is more important than LB because $P_A \cong UB$ can provide higher molecular communication capacity. Therefore, in the numerical analysis in Section 6 we evaluate only UB for the selection of P_A.

5.2   An Adaptive Molecular Error Compensation Scheme

In the molecular communication, error rate is heavily affected by some environmental factors such as binding rate (k_1), temperature (T), distance (α) between nanomachines. Therefore, it is imperative to compensate the errors due to the changing environmental factors to achieve higher molecular communication capacity. For this compensation, it is essential to regulate molecular bit transmission probability (P_A) with respect to the changing environmental factors. However, the regulation of P_A to compensate the errors needs some coordination between TN and RN. This coordination enables an adaptive error compensation scheme that is periodically conducted by TN and RN to compensate possible channel errors. Here, we introduce an adaptive Molecular Error Compensation (MEC) scheme for the molecular communication between TN and RN as outlined below:

1. Initially, TN sets the molecular bit transmission probability P_A to an initial value denoted by $\overline{P_A}$ and initiates the molecular communication.
2. When error rate increases, RN detects the increasing error rate[4].
3. RN emits a molecular bit stream denoted by BS_1 to terminate current molecular communication between TN and RN and to initiate the error compensation scheme. BS_1 is a special stream[5] such that when it is received by TN, TN can immediately terminate the current molecular communication and infer the initiation of the error compensation scheme. Therefore, by means of BS_1, RN can initiate the error compensation scheme in any time when it detects increasing error rate in the molecular channel.
4. Once TN receives BS_1, it immediately emits the molecular bit stream BS_2 [6] like $A00000A00000$ to enable RN to compute N_{ex} and UB.
5. Using N_{ex} and UB, RN selects P_A as a value closest to UB ($P_A \cong UB$).
6. RN informs TN about the selected molecular bit transmission probability. Here, we do not assume that RN communicates the actual value of the selected P_A, which is possibly a floating point number, to TN. We only assume that RN emits specific molecular bit patterns corresponding to the different level of molecular bit transmission probability such that according

---

[4] Note that while it may be possible to develop some error detection mechanisms for molecular communication, it is beyond the scope of this paper.

[5] BS_1 is a fixed molecular bit stream, which may be determined in the design stage of the molecular communication system. For example, $A0A$ may be selected as BS_1.

[6] Since N_{ex} is the number of molecules delivered within an exponential decaying phase after TN emits A, BS_2 is appropriate to enable RN to compute N_{ex}. Furthermore, other molecular bit streams that can enable this computation can be selected for this computation.

Algorithm 1. MEC

---

1 TN sets P_A as $\overline{P_A}$
2 TN initiates the molecular communication
3 **foreach** P_A **do**
4   RN detects the increasing error rate
5   RN emits BS_1
6   TN terminates the molecular communication
7   TN emits BS_2
8   RN computes N_{ex} and UB
9   RN selects P_A, $(P_A \cong UB)$
10   RN informs TN about the selected P_A
11   TN updates P_A as the selected P_A
12   TN emits BS_3
13   TN again initiates the molecular communication
14 **end**

---

to the emitted molecular bit pattern, TN infers the selected molecular bit transmission probability.

7. TN sets P_A as the selected molecular bit transmission probability. After the setting of the molecular bit transmission probability, TN emits the molecular bit stream BS_3[7] to again initiate the molecular communication. Then, TN again initiates the molecular communication according to the updated molecular bit transmission probability, which minimizes the error rate and maximizes the molecular communication capacity.

Using the adaptive Molecular Error Compensation (MEC) scheme given above, TN and RN can collaboratively select P_A to minimize the error rate and maximize molecular communication capacity. We also give MEC in the pseudo-code given in Algorithm 1.

6   Numerical Analysis

In this section, we first present the numerical analysis performed over the mutual information expression given in (17) to show how the molecular communication capacity varies according to the some environmental parameters and some other parameters specific to the nanomachines TN and RN. Then, we give the numerical analysis for the performance of MEC over the selection of the most appropriate P_A that provides high molecular communication capacity with minimum error rate. The aim of this analysis is to determine appropriate configuration of molecular communication parameters, which can achieve high molecular communication capacity, according to changing environmental factors. We perform the numerical analysis using Matlab. We assume that TN and RN are randomly positioned in an environment, which may have different diffusion coefficients

---

[7] Similar to BS_1, BS_3 is also a fixed molecular bit stream, which determined in the design. For example, $0A0$ may be selected as BS_3.

Table 1. Simulation Parameters

Binding rate (k_1)	0.1-0.5 $(\mu mol/liter/s)$
Zero-force release rate (k^0_{-1})	0.08 (s^{-1})
Temperature (T)	300-1000 K
Distance between TN and RN (α)	$5^{-10} - 4 \times 10^{-9} m$
Applied force per bound (f)	10^{-12} (J/m)
Concentration of molecules A (L_{ex})	1-8 $(\mu mol/liter)$
Duration of the pulses (t_H)	0.5-1 s
Number of receptors R (N)	0.001-0.01 $(\mu mol/liter)$
S $(\mu mol/liter)$	$10^{-6} - 4 \times 10^{-6}$

such that it allows TN to achieve different binding rates (k_1). Due to the principles of mass action kinetics, we also assume that k_1 varies with temperature of environment (T) and distance (α) between TN and RN such that k_1 is directly proportional with $2T$ $(k_1 \propto 2T)$ and inversely proportional with α $(k_1 \propto 1/\alpha)$, respectively. Moreover, we assume that k^0_{-1} depends only on the properties of RN receptors and cannot be changed. The simulation parameters of the analysis are given in Table 1.

6.1 Effect of Environmental Factors on Molecular Communication Capacity

Binding Rate: For the first analysis, we investigate the effect of binding rate (k_1) on capacity of the molecular channel. In Fig. 2, mutual information (M) given in (17) is shown with varying molecular bit transmission probability (P_A) for different k_1. For higher $k_1 = 0.2 - 0.4$ $\mu mol/liter/s$, TN delivers higher number of molecules A that is greater than S. Therefore, for higher k_1, transmission of molecular bit 0 can be erroneous and molecular communication capacity decreases. However, for a smaller $k_1 = 0.1$ $\mu mol/liter/s$ that enables TN to deliver sufficient molecules for molecular bit A and 0, M can be maximized selecting an appropriate P_A. TN can deliver greater than S number of molecules in transmission of molecular bit A and deliver less than S number of molecules in transmission of molecular bit 0. Therefore, to achieve high molecular communication capacity, it is necessary to select appropriate S with respect to binding rate (k_1). For an environment imposing higher k_1, S should be selected as a sufficiently high value that can hinder the delivery of erroneous molecular bit 0. For an environment imposing smaller k_1, a smaller S should be used such that it can hinder the delivery of erroneous molecular bit A.

Temperature: Temperature of the environment (T) is another important parameter since it heavily affects the binding rate (k_1) and molecular communication capacity. However, T has similar effects with k_1 on the molecular communication capacity. In Fig. 3, M is shown with varying P_A for different T. For $T = 300 - 500$ K, higher molecular communication capacity can be achieved. However, the capacity decreases while T is further increased from 500 K to 1000 K $(T = 500 - 1000$ $K)$. This stems from binding rate (k_1) that increases

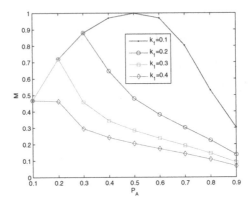

Fig. 2. M with varying P_A for different k_1

with the higher temperature (T) such that the increasing k_1 results in delivery of excessively high number of molecules and erroneous molecular bit 0 and the capacity decreases. Hence, for an environment having higher T, S should be selected as a sufficiently high value to prevent erroneous molecular bit 0 and to maximize molecular communication capacity.

Distance between TN and RN: As the traditional wireless communication, distance (α) between TN and RN heavily affects molecular communication capacity since k_1 depends on α. While α decreases, k_1 increases. To evaluate the effect of α on molecular communication, in Fig. 4, M is shown with varying P_A for different α. For smaller $\alpha = 5 \times 10^{-10} - 20 \times 10^{-10}$ m, k_1 increases and excessive number of molecules that is greater than S is delivered to RN. This results in erroneous molecular bit 0 and decreases the capacity. However, for a sufficiently high $\alpha = 40 \times 10^{-10}$ m providing appropriate k_1, TN delivers a number of molecules less than S in transmission of molecular bit 0 and delivers greater than S number of molecules in transmission of molecular bit A. This can maximize the molecular communication capacity. Therefore, S should be selected with respect to distance between TN and RN. While the distance increases, S should be decreased to prevent erroneous molecular bit A. While the distance decreases, S should be increased to prevent erroneous molecular bit 0.

6.2 Effect of the Parameters Specific to TN and RN on Molecular Communication

In this section, we present the numerical results for effect of some parameters specific to TN and RN. S is one of the most important molecular communication parameter specific to TN and RN. In Fig. 5(a), M is shown with varying P_A for different S. For smaller $S = 1 \times 10^{-6} - 2 \times 10^{-6}$ $\mu mol/liter$, it is most likely that TN can deliver a number of molecules A that is greater than S. Therefore, molecular communication capacity decreases while S decreases. However, using a sufficiently high $S = 3 \times 10^{-6} - 4 \times 10^{-6}$ $\mu mol/liter$, which enables non-erroneous

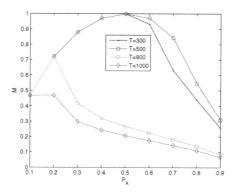

Fig. 3. M with varying P_A for different T

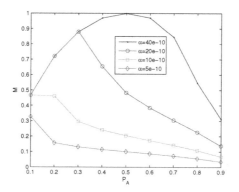

Fig. 4. M with varying P_A for different α

molecular communication bits, maximum molecular communication capacity can be achieved. Hence, S should be selected according to some environmental factors such as binding rate (k_1), temperature (T), and distance between TN and RN (α). For an environment imposing high k_1 and T, S should be used as a smaller value to prevent erroneous molecular bit 0.

Concentration of emitted molecules (L_{ex}) is also one of the most important parameters specific to TN and RN, which affects the number of delivered molecules in each transmission of molecular bits. In Fig. 5(b), M is shown with varying P_A for different L_{ex}. For $L_{ex} = 1$ $\mu mol/liter$, L_{ex} is sufficiently high such that TN can achieve to deliver the needed concentration to RN for molecular bits A and 0. Therefore, high molecular communication capacity can be achieved and they can be maximized using appropriate P_A. However, for $L_{ex} = 2 - 8$ $\mu mol/liter$, TN delivers excessive concentration to RN for molecular bits 0, which is greater than S and erroneous bits arise and the molecular communication capacity decreases. Therefore, to achieve higher molecular communication capacity, L_{ex} must be selected as an appropriate value according to the environmental factors such as binding rate, temperature and distance between TN and RN.

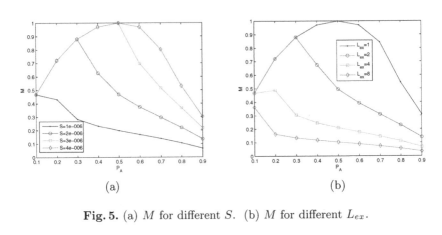

Fig. 5. (a) M for different S. (b) M for different L_{ex}.

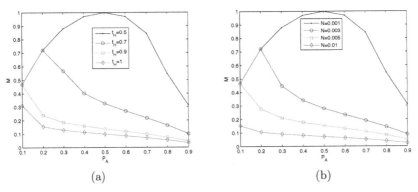

Fig. 6. (a) M for different t_H. (b) M for different N.

Duration of emission pulse (t_H) is critical for performance of the molecular communication. While t_H increases, number of delivered molecules increases. Therefore, erroneous molecular bits 0 arise while t_H increases. In Fig. 6(a), M is shown with varying P_A for different t_H. For $t_H = 0.5\ s$, high molecular communication capacity can be achieved. However, the capacity decreases at higher values of P_A while t_H increases. As t_H and P_A increase, number of delivered molecules increases such that TN cannot deliver the concentration smaller than S in transmission of molecular bit 0. Therefore, erroneous molecular bit 0 arises at higher P_A while t_H increases. Hence, appropriate t_H is needed to achieve higher molecular communication capacity.

Concentration of receptors (N) on RN also affects the number of delivered molecules in the molecular communication. While N increases, number of delivered molecules increases. In Fig. 6(b), M is shown with varying P_A for different N. For $N = 0.003 - 0.01\ \mu mol/liter$, TN delivers excessive number of molecules in transmission of molecular bit 0 and erroneous molecular bits arise and the molecular communication capacity decreases. For $N = 0.001\ \mu mol/liter$, the concentration of receptors on RN is sufficient to enable TN to deliver sufficient molecular concentration for molecular bit A and 0. Therefore, selecting appropriate N, the molecular communication capacity can be maximized. When the environment allows delivery of smaller number of molecules, N should be selected as a higher value to enable TN to deliver sufficient number of molecules for non-erroneous molecular bit A.

6.3   Adaptive Molecular Error Compensation Scheme

In this section, we present the numerical analysis on performance of adaptive Molecular Error Compensation (MEC) scheme. For the performance of MEC, we show the mutual information given in (17) with and without MEC scheme according to the varying environmental factors with higher error rate. For the case with MEC, we allow MEC scheme to select molecular transmission probability (P_A) to compensate possible molecular errors and to achieve high molecular communication capacity according to varying environmental factors. For the case without MEC, we statically set P_A as $P_A = 0.5$ regardless of changing environmental factors. We show the effect of MEC on the molecular communication capacity in terms of variation of three environmental factors, binding rate (k_1), temperature (T), and distance between TN and RN (α). Throughout the analysis, MEC sets molecular transmission probability (P_A) as $P_A = \lfloor 10UB \rfloor /10$, where $\lfloor . \rfloor$ denotes the floor function. This selection strategy provides an appropriate P_A for MEC such that $P_A \cong UB$.

Binding Rate: In Fig. 7, mutual information (M) given in (17) is shown for varying binding rate (k_1) with and without MEC. As k_1 increases ($k_1 = 0.1 - 0.5 \ \mu mol/liter/s$), molecular communication capacity decreases due to erroneous molecular bit 0. However, MEC can compensate some possible errors selecting the most appropriate P_A. MEC provides % 100 capacity improvement with repect to the case that statically selects $P_A = 0.5$ without MEC.

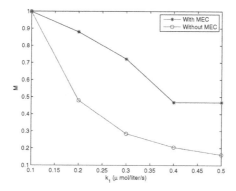

Fig. 7. M for varying k_1 with and without MEC

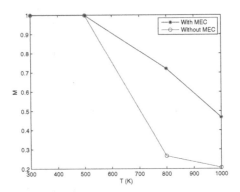

Fig. 8. M for varying T with and without MEC

Temperature: In Fig. 8, M is shown for varying temperature (T) with and without MEC. While T increases, the molecular communication capacity decreases due to increasing error rate in transmission of molecular bit 0. For $T = 300 - 500\ K$, almost the same capacity can be achieved with and without MEC because $\lfloor 10UB \rfloor /10 \cong 0.5$. However, as T further increases, MEC significantly overcomes the static selection strategy in which P_A is set as $P_A = 0.5$ selecting the most appropriate P_A and MEC can compensate possible errors and achieve high molecular communication capacity.

Distance between TN and RN: In Fig. 9, M is shown for varying distance between TN and RN (α). As α decreases, the molecular communication capacity decreases due to increasing error rate in transmission of molecular bit A. However, MEC can compensate these errors and achieve high molecular communication capacity selecting the most appropriate P_A. MEC provides more than $\%\ 100$ capacity imporvement with respect to the static selection strategy using $P_A = 0.5$.

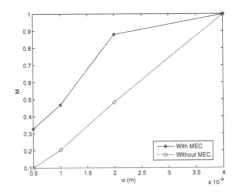

Fig. 9. M for varying α with and without MEC

7   Conclusion

In this paper, we derive a closed form expression for capacity of the channel between TN and RN. Furthermore, we propose an adaptive Molecular Error Compensation (MEC) scheme for the molecular communication between TN and RN. MEC allows TN and RN to collaboratively select the most appropriate appropriate molecular bit transmission probability to maximize molecular communication capacity with respect to environmental factors such as temperature, binding rate, distance between nanomachines. Using the capacity expression, we investigate how the conditions such as temperature of environment, concentration of emitted molecules, distance between nanomachines and duration of molecule emission, binding rate, concentration of receptors affect the molecular communication capacity. Numerical analysis reveals that MEC provides more than % 100 capacity improvement in the molecular communication selecting the most appropriate molecular transmission probability that proactively compensate the possible errors in the molecular channel. Numerical analysis also shows that the molecular communication with high capacity is only possible by arranging the molecular communication parameters such that cross-relation between the parameters should be carefully considered to compensate their negative effects over each other. Furthermore, a possible design scheme for the molecular communication should consider the environmental factors to provide high molecular communication capacity. The design scheme should select the parameters specific to TN and RN according to the environmental factors.

Acknowledgments. This work was supported in part by TUBA-GEBIP Programme and TUBITAK under the Grant #106E179.

References

1. Hiyama, S., Moritani, Y., Suda, T., Egashira, R., Enomoto, A., Moore, M., Nakano, T.: Molecular Communication. In: NSTI Nanotech, Anaheim, California, USA, pp. 391–394 (2005)
2. Whitesides, G.M.: The Once and Future Nanomachine. Scientific American 285(3), 78–83 (2001)
3. Suda, T., Moore, M., Nakano, T., Egashira, R., Enomoto, A.: Exploratory Research on Molecular Communication between Nanomachines. In: Genetic and Evolutionary Computation Conference (GECCO), Washington, DC, USA (2005)
4. Moore, M., Enomoto, A., Nakano, T., Egashira, R., Suda, T., Kayasuga, A., Kojima, H., Sakakibara, H., Oiwa, K.: A Design of a Molecular Communication System for Nanomachines Using Molecular Motors. In: IEEE PERCOMW, Italy, pp. 554–559 (2006)
5. Moritani, Y., Hiyama, S., Suda, T.: Molecular Communication for Health Care Applications. In: IEEE PERCOMW 2006, Italy, pp. 549–553 (2006)
6. Nakano, T., Suda, T., Moore, M., Egashira, R., Enomoto, A., Arima, K.: Molecular Communication for Nanomachines Using Intercellular Calcium Signaling. In: IEEE Conference on Nanotechnology, Nagoya, Japan, pp. 478–481 (2005)

7. Hiyama, S., Isogawa, Y., Suda, T., Moritani, Y., Sutoh, K.: A Design of an Autonomous Molecule Loading/Transporting/Unloading System Using DNA Hybridization and Biomolecular Linear Motors. In: European Nano Systems, Paris, France, pp. 75–80 (2005)

8. Atakan, B., Akan, O.B.: An Information Theoretical Approach for Molecular Communication. In: ACM BIONETICS 2007, Budapest, Hungary (2007)

9. Rospars, J.P., Krivan, V., Lansky, P.: Perireceptor and receptor events in olfaction. Comparison of concentration and flux detectors: a modeling study. Chem. Sens. 25, 293–311 (2000)

10. Krivan, V., Lansky, P., Rospars, J.P.: Coding of periodic pulse stimulation in chemoreceptors. Elsevier Biosystem 67, 121–128 (2002)

11. Saxton, M.J.: Anomalous Diffusion Due to Binding: A Monte Carlo Study. Biophysical Journal 70, 1250–1262 (1996)

12. Long, M., Lü, S., Sun, G.: Kinetics of Receptor-Ligand Interactions in Immune Responses. Cell. & Mol. Immuno. 3(2), 79–86 (2006)

13. Bell, G.I.: Models for the specific adhesion of cells to cells. Sciences 200, 618–627 (1978)

14. Camacho, C.J., Kimura, S.R., DeLisi, C., Vajda, S.: Kinetics of Desolvation-Mediated Protein Binding. Biophysical Journal 78, 1094–1105 (2000)

15. Cover, T.M., Thomas, J.A.: Elements of information theory. John Wiley-Sons, Chichester (2006)

Molecular Communication through Gap Junction Channels

Tadashi Nakano[1], Tatsuya Suda[1], Takako Koujin[2],
Tokuko Haraguchi[2], and Yasushi Hiraoka[2]

[1] Department of Computer Science
Donald Bren School of Information and Computer Sciences
University of California, Irvine
{tnakano,suda}@ics.uci.edu
[2] Kobe Advanced ICT Research Center
National Institute of Information and Communications Technology
{koujin,tokuko,yasushi}@nict.go.jp

Abstract. Molecular communication is engineered biological communication that allows biological devices to communicate through chemical signals. Since biological devices are made of biological materials and are not amenable to traditional communication means (e.g., electromagnetic waves), molecular communication provides a mechanism for biological devices to communicate by transmitting, propagating, and receiving molecules that represent information. In this paper, we explore biological cells and their communication mechanisms for designing and engineering synthetic molecular communication systems. The paper first discusses the characteristics and potential design of communication mechanisms, and then reports our experimental and modeling studies to address physical layer issues of molecular communication.

1 Introduction

Molecular communication [18, 33] plays a critical role in a broad range of biological applications from molecular computing, biochemical sensing to nanomedicine. Molecular communication provides a means by which system components coordinately perform large-scale complex tasks that can not be accomplished by individual components. For example, engineered organisms that function as basic logic gates can perform distributed computing through communication [45]; medical nanomachines with communication capabilities can perform coordinated monitoring of human health [15].

It is well known that living cells in various tissues and organisms utilize numerous networking mechanisms to establish cell-cell communications. Existing cell-cell communication mechanisms may be therefore applicable to engineering of synthetic molecular communication systems. In addition, current molecular engineering technology may allow modification of existing cell-cell communications to provide advanced functionality to meet various needs of biological applications.

C. Priami et al. (Eds.): Trans. on Comput. Syst. Biol. X, LNBI 5410, pp. 81–99, 2008.

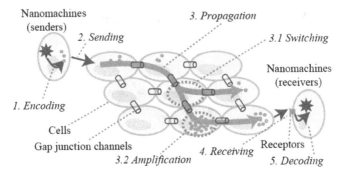

Fig. 1. Molecular Communication through Gap Junction Channels

In our early work [34, 35][1], we designed a class of molecular communication inspired by cell-cell communications through gap junction channels (Figure 1). Gap junction channels are intercellular channels made of proteins that allow biological cells to exchange molecules with attached cells [38]. As in naturally occurring biological cellular networks, gap junction channels in our design are used to interconnect biological cells, providing a cell-cell communication medium for nanomachines. Nanomachines are engineered organisms or biological devices that are programmed to achieve application specific goals, and chemically communicate over the cell-cell communication medium. To support molecular communication between nanomachines, the communication medium may provide various networking mechanisms such as signal amplification and switching.

To illustrate how molecular communication is performed and what communication processes are generally involved, consider the following example. The sender senses some toxic chemical substances in the environment and synthesizes stimulating molecules that can potentially initiate cell-cell signaling (encoding). The sender then emits the stimulating molecules (sending), which trigger cell-cell signaling. Upon receiving the stimulating molecules, cells generate diffusive signal molecules and propagate them through gap junction channels intercellularly (propagation). During propagation, signal molecules may be directed toward target receivers (switching) or amplified for longer distance propagation (amplification). The receiver reacts to nearby cells that start receiving the signal molecules (receiving), and initiates biochemical processes such as secreting neutralizing chemicals into the environment (decoding).

The rest of the paper is organized as follows. Section 2 describes general characteristics and features of molecular communication systems. Section 3 discusses a possible design of networking mechanisms based on gap junctional communication. Section 4 experimentally demonstrates our first prototype of a gap junction based molecular communication medium (called a cell wire). Section 5 describes a modeling study performed to understand communication related characteristics of molecular communication through gap junction channels and Section 6 concludes this paper.

---

[1] A part of the paper was presented at [35]. This paper performs a further modeling study to investigate additional physical layer issues of calcium signaling through gap junction channels.

2 System Characteristics

Molecular communication systems may exhibit several distinct features from silicon based counterparts such as computer communication systems. As will be described in the following, such features may be explored to design and develop new biological ICT (Information and Communications Technology) applications (e.g., biocompatible implant systems) while at the same time such features may limit the applicability of molecular communication.

Small scale, limited range and slow speed communication: The system size may vary from um to cm depending on cell types (10~30um for eukaryotic cells) and a cellular structure (cm or more) formed to implement a system. The communication range between system components (cells) is strictly limited and communication speed is extremely slow compared to existing telecommunications (i.e., speed of light). For example, the longest range and fastest communication possible would be when neural signaling is used as a communication channel, in which case electro-chemical signals (action potentials) may propagate up to several meters at 100 m/sec. In other cases, where Ca^{2+} waves of astrocytes are used, communication is much slower and 20 um/sec within a cm range [38].

Functional complexity: An advantageous feature of using cells as a system component is achievable functional complexity. A cell is a highly functional and integrated component with information processing capabilities. A cell has a number of sensors (e.g., receptors to sense the environment), logic circuits (e.g., complex signal transduction pathways), memory for storing information, and actuators that can generate motion. A functional density of a bacterial species, *Escherichia coli*, is estimated in [11], which states that a cell stores a 4.6 million basepair chromosome in a 2 um^2 area, which is equivalent to a 9.2-megabit memory that encodes a number of functional polypeptides. Functional complexity may help a design of highly compacted engineered systems including NEMS/MEMS (nano- and micro-electromechanical systems), lab-on-a-chips and u-TAS (Micro-Total Analysis Systems).

Biocompatibility: Another advantageous feature of using cells is biocompatibility. Cells can interact directly with other cells, tissues and organs through receiving, interpreting, synthesizing and releasing molecules, and thus molecular communication systems may be useful in medical domains (e.g., implantable devices, cell-based biosensors, body sensor networks [46]), where interactions with a human body are necessary. Also, molecular communication may be indispensable for communication between soft nanomachines that are composed of biological materials (e.g., molecular computing devices) and that are not capable of transmitting and receiving traditional communication signals (electromagnetic waves).

Chemical energy, energy efficiency, and low heat dissipation: Molecular communication systems operate with chemical energy (e.g., ATP), unlike silicon devices that require electric batteries. Chemical energy may be possibly supplied by the environment where molecular communication systems are situated. For example, molecular communication systems deployed in a human body may harvest energy (e.g., glucose) from the human body, requiring no external energy sources. Also, molecular communication systems may be energy efficient with low heat dissipation as cellular components are energy efficient. For example, an F1-ATP motor converts ATP energy to mechanical work at nearly 100 percent energy efficiency [22].

Self-assembly: Self-assembly is a possible property of molecular communication systems. Cells can divide and grow to assemble into a larger structure (organs). This self-assembly property may be exploited in system design of molecular communication systems, enabling a bottom up system fabrication and deployment. In addition, self-assembling molecular communication systems are highly fault-tolerant as damaged parts of a system may be recovered through division and growth of nearby cells.

Probabilistic behavior: Molecular communication systems will be placed in an aqueous environment where thermal noise and other environmental factors may affect the system behavior. For example, signal molecules may randomly propagate based on Brownian motion. Signal molecules may also be broken down or degraded during propagation, introducing probabilistic and unpredictable behavior. For molecular communication, such probabilistic aspects may be overcome by relying on a large number of molecules for communication (ensemble averaging). On the other hand, thermal noise may be utilized to enhance a signal-to-noise ratio with stochastic resonance as is theoretically presented in neural information processing [27].

Robustness and fragility: Cells or cellular systems exhibit some degree of robustness against internal and external perturbations [23]. Cells have evolutionary acquired control mechanisms (feedback mechanisms) to achieve robustness and maintain stability (e.g., homeostasis). At the same time, cells are extremely fragile to various factors such as temperature and pH changes that can destruct system behavior easily. Molecular communication systems may inherit such features of robustness and fragility.

Safety issues: Although molecular communication systems have a number of features that silicon devices may not posses, safely issues definitely arise, especially when used for medical and environmental applications (e.g., that may introduce new infectious viruses). Potential risks must be assessed in order to advance this technology field.

3 System Design

Synthetic biological communication systems have been designed based on naturally exiting cell-cell communication mechanisms and experimentally demonstrated in the synthetic biology area. For example, cell-cell communications and intracellular signal processing mechanisms are modified to build spatial patterns of cells [4] and to regulate population of cells [47]. These systems use unguided communication media in which bacterial cells diffuse paracrine type signals in the extracellular environment. Another type of communication is possible by using guided communication media such as gap junction channels that can guide diffusion of signal molecules.

    In this paper, we explore design and engineering of the latter type of molecular communication, specifically calcium signaling through gap junction channels. Calcium signaling through gap junction channels is one of the most common forms of cell-cell communications in mammalian cells, and a brief and necessary introduction is first provided in the following subsections. Our initial ideas on design of molecular communication based on calcium signaling through gap junction channels are also presented.

3.1 Gap Junction Channels

Gap junction channels [38] are physical channels formed between two adjacent cells, connecting the cytoplasm of the two cells (Figure 2). A gap junction channel consists of two apposed hexamers of connexin proteins around a central pore. There are over 20 connexins reported, and different connexins can constitute gap junction channels with different properties in terms of permeability and selectivity of molecules [14]. Gap junction channels normally allow the passage of small molecules (<1000Da) (e.g., ions, IP$_3$, cyclic AMP) between cells. Also, gap junction channels can vary the permeability and selectivity in response to various internal and external factors including cytosolic Ca^{2+} concentrations, membrane potential, connexin phosphorylation/dephosphorylation, electromagnetic fields, temperature and pH in the environment.

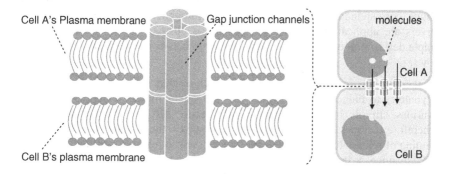

Fig. 2. Gap junction channels are cell-cell communication channels formed between two adjacent cells, allowing small molecules (i.e., ions and metabolites) to diffuse between the cytoplasms of the two cells

Gap junction channels play an important role in propagating signal molecules including Ca^{2+} signals, known as intercellular Ca^{2+} wave propagation. For example, glial cells have 30-150 nM of cytosolic Ca^{2+} concentration (denoted as [Ca^{2+}]$_i$) at rest, and upon stimulated, [Ca^{2+}]$_i$ increases up to hundreds of nM or several μM within milliseconds while propagating the increase from cell to cell [9].

Intercellular Ca^{2+} waves, mediated by gap junctional communication, can potentially propagate biological information such as cell death [26] and growth [41]. More complex information may be propagated because Ca^{2+} signals are used as a universal second messenger in diverse cell types. It is known that temporal and spatial dynamics of cytosolic Ca^{2+} (termed as Ca^{2+} spikes/oscillations) are modulated to encode various cellular activities, and the Ca^{2+} dynamics is decoded by Ca^{2+} sensitive molecules that trigger various cellular responses [7, 8].

3.2 Molecular Communication through Gap Junction Channels

Molecular communication through gap junction channels is described below relative to the generic molecular communication architecture proposed in [33]. The generic architecture is composed of *signal molecules* that carry information to be transmitted, *senders* that transmit signal molecules, *receivers* that receive signal molecules, and the *environment* in which signal molecules propagate from senders to receivers. Molecular communication in the generic architecture is generalized by the five key communication processes: *encoding* (process by which senders translate biochemical reactions into signal molecules), *sending* (process by which senders release signal molecules into the environment), *propagation* (process by which signal molecules move through the environment from senders to receivers), *receiving* (process by which receivers detect signal molecules propagating in the environment), and *decoding* (process by which receivers, after detecting signal molecules, decodes the molecules into biochemical reactions.)

In the present molecular communication through gap junction channels, Ca^{2+} signals are used as signal molecules. Ca^{2+} signals are universal second messengers capable of propagating within a cell and between cells, and its dynamics are used to represent information in biological systems (cf. section 3.1). In the present molecular communication, gap junctionally connected cells serve as the communication medium, corresponding to the environment in the generic architecture. Senders and receivers are application specific and remain as abstract entities in this paper. An example description of molecular communication between senders and receivers is as follows.

1. **Encoding**: Encoding is the process by which senders select the properties of Ca^{2+} dynamics to induce and propagate. Ca^{2+} dynamics may be externally induced by various stimuli such as agonistic substances. An example encoding process therefore includes choices of agonistic substances, the amounts to release, and the timing of release (e.g., agonistic substances may be released constantly, in a burst-like manner, or in an even more complex manner.)
2. **Sending:** Sending is the process of releasing agonistic substances in the manner decided in the encoding process. The released agonistic substances react with membrane receptors of neighboring cells, activating downstream signaling pathways that lead to increase of $[Ca^{2+}]_i$ in the cells.
3. **Propagation**: The increase of $[Ca^{2+}]_i$ propagates from cell to cell through gap junction channels. (It can be propagated through various molecular mechanisms as reported in Chapter 7 of [38].) Cells may perform amplification of cytosolic Ca^{2+} signals to prevent signal level attenuation using calcium induced calcium release. Cells may also perform switching by opening and closing gap junction channels, so that Ca^{2+} signals propagate preferentially toward specific receivers.
4. **Receiving**: In receiving, receivers detect cellular responses of neighboring cells that have received propagating Ca^{2+} signals. For example, neighboring cells experiencing changes in its $[Ca^{2+}]_i$ may release molecules into the extracellular environment, and nearby receivers capture the released molecules using membrane receptors.
5. **Decoding**: Receivers invoke application specific responses corresponding to molecules captured.

3.3 Design Detail

Some more detail of the system design follows in comparison to computer networks design [42]. Looking into computer networks design may help identify necessary mechanisms and functionality for molecular communication. For example, typical physical layer issues such as types of communication media, modulation techniques, signal amplification and noise handling may also be important for molecular communication. In addition, mechanisms and functionality from upper layers of computer networks may also be useful for molecular communication, including synchronization, error detection/ correction, filtering, switching, addressing, flow control, and so on. A possible design of some of the mechanisms and functionality is discussed below.

Communication media: There are two common types of communication media by which cells communicate with each other. In one type, similar to guided media of computer networks (e.g., coax cables, fiber optics), cells establish direct contact through gap junction channels, restricting diffusion of signal molecules (e.g., Ca^{2+}, IP_3) within cells connected through gap junction channels. In the other type, similar to unguided media of computer networks (e.g., wireless), cells diffuse signal molecules (e.g., ATP, cyclic AMP) in the extracellular environment, and nearby cells respond to the molecules in the environment (paracrine signaling). Accordingly, the present molecular communication can also be performed through the two types of communication media.

Modulation techniques: Cells adopt various modulation schemes to represent data (or messages), similar to modulation technique used in computer networks or radio broadcast (FM/AM). For example, in calcium signaling, messages are represented in complex temporal dynamics of calcium concentrations, referred to as AM and FM of calcium signaling [6]. Similarly, some cells can secrete molecules in the extracellular environment in a complex manner (e.g., in an oscillatory manner) to represent some messages. Accordingly, the present molecular communication may be able to utilize their built-in modulation techniques.

Signal attenuation and noise handling: Cells use various feedback mechanisms; positive feedback to amplify signals and negative feedback to reduce the impact of noise. For example, CICR (Calcium Induced Calcium Release) is a mechanism to amplify calcium signals. It is recently found that ATP signals are also amplified in some cell types termed as ATP induced ATP release [2]. For noise handling, there is theoretical work that thermal noise is utilized by cells to increase a signal-to-noise ration through SR (Stochastic Resonance) [27]. These mechanisms may also be incorporated into the present molecular communication.

Filtering: Gap junction channels can have different selectivity and permeability, and this property can be used to implement filters for molecular communication, similar to packet filtering of computer networks. Figure 3 (a) illustrates a simple example of filtering using gap junction channels, where gap junction channels formed between cells have different permeability and thus signal molecules preferentially propagate.

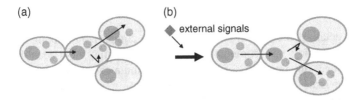

Fig. 3. The four cells together perform signal filtering and switching. (a) Incoming signal molecules represented as circles is more permeable to gap junction channels as indicated by the arrows; the signal molecules propagate toward the direction while those propagating toward the other direction are filtered out. (b) The external signal represented by the rectangle modifies permeability of gap junction channels, dynamically switching the direction of signal propagation.

Switching: The selectivity and permeability of gap junction channels can vary in response to various factors (e.g., whether connexin proteins are phosphorylated), and this dynamic property can be used to implement dynamic switching mechanisms. Figure 3 (b) illustrates a design of dynamic switching, where the permeability to signal molecules are changed by external signals.

Signal aggregation: The selectivity and permeability of gap junction channels can be further exploited to perform more complex networking functionality. Figure 4 shows an example of signal aggregation using two different types of signals. One type of signals flows into the centered cell, but this type of signals is not permeable to the rightmost cell, therefore no further propagation. Once the other type of signals comes in the centered cell, the two types of signals chemically react to produce new signals that are permeable to the rightmost cell, therefore propagating signals.

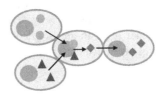

Fig. 4. The four cells together perform signal aggregation. Two types of signal molecules incoming to the centered cell, represented as circles and triangles, are both not permeable to the rightmost cell. However, the two types of signals react in the centered cell to produce signal molecules represented as rectangles that are permeable to the rightmost cell, propagating signals to the rightmost cell only when two types of signals (represented as circles and triangles) come in the centered cell.

4 Experimental Demonstration

HeLa cells (Human epithelial cells) that were deficient in gap junctional communication have been genetically engineered to express functional gap junction channels in [14], and patterned into a straight line to demonstrate a simple molecular commutation medium called *cell wires* [36]. Cell wires propagate Ca^{2+} waves along a line of gap junction transfected HeLa cells (HeLa Cx43 cells). The representative experimental results are shown in Figure 5. Briefly, a microplatform for cell-patterning was first established by utilizing lithography and surface chemical treatment, and HeLa Cx43 cells were patterned onto the platform with predefined geometry. Flash-photolysis of caged-ATP was then used to initiate intercellular Ca^{2+} waves; cell A was flashed, which increased the Ca^{2+} level. The increased Ca^{2+} level propagated along the straight line about 5 um/sec, and reached the cell that was 10 cells away from the flashed cell (cell A). The experimental results were reproducible for a number of times ($n \geq 10$), and similar results were obtained from repeated experiments.

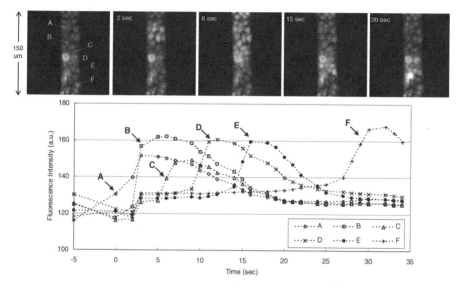

Fig. 5. Cell wires demonstrated in [36]. (Top) Cell A was flashed using flash-photolysis of caged-ATP. A Ca^{2+} wave was generated at cell A and propagated cell-to-cell along the line of patterned cells. The five images shown were obtained through fluorescence microscopy before flash photolysis and 2, 8, 15, 20 seconds after the flash photolysis. (Bottom) The fluorescence intensity of cells A-F during the fluorescence imaging is plotted.

The Ca^{2+} wave propagation in our experimental setup was presumably facilitated by gap-junctional diffusion of messenger molecules (e.g., Ca^{2+} itself, IP_3) in addition to extracellular diffusion of photo-released ATP. It is noted however that intercellular Ca^{2+} wave can propagate through different molecular mechanisms [38], and further experiments need to be conducted to identify the mode of Ca^{2+} wave propagation observed in our experiments.

The achieved distance and speed of Ca^{2+} wave propagation, although within the commonly observed range for other cell types that are naturally capable of gap junctional communication (i.e., a few um to 500 um or more, at a few um to 100 um/sec [38]), may be unsatisfactory for specific needs of biological applications. Increasing the possible distance and speed of communication may benefit a wider range of applications. Gap junctional connectivity is one of the factors influencing the property of communication, and how gap junctional connectivity affects the speed and distance of propagating Ca^{2+} waves is examined in the following modeling study.

5 Theoretical Prediction

The following modeling study investigates physical layer issues of calcium signaling through gap junction channels. A mathematical model is used to examine the impact of gap junctional coupling on the properties of intercellular Ca^{2+} waves. Figure 6 schematically illustrates the model used in this study. The model is based on a classical model of calcium oscillation of a single cell[2] [16], and a minor extension is made to simulate intercellular Ca^{2+} waves over an array of cells. Briefly, the Ca^{2+} concentration of the cytosol at cell i is represented as Z_i and that of the calcium store is Y_i ($i=0,1,2,\cdots,n-1$). The time evolution of Z_i and Y_i is described as follows:

$$\frac{dZ_i}{dt} = v_0 + in_i - v_{2i} + v_{3i} + k_f Y_i - kZ_i + P(Z_{i-1} - 2Z_i + Z_{i+1})$$

$$\frac{dY_i}{dt} = v_{2i} - v_{3i} - k_f Y_i$$

$$v_{2i} = V_{M2} \frac{Z_i^{\,n}}{K_2^{\,n} + Z_i^{\,n}}$$

$$v_{3i} = V_{M3} \frac{Y_i^{\,m}}{K_R^{\,m} + Y_i^{\,m}} \frac{Z_i^{\,p}}{K_A^{\,p} + Z_i^{\,p}}$$

Most of the constants and notations in the above equations are barrowed from [16]; v_0 represents Ca^{2+} influx from the extracellular environment to the cytosol; in_i a stimuli applied to cell i (replaced with $v_1 \cdot B$ in [16]); v_{2i} and v_{3i} respectively Ca^{2+} uptake to and release from the calcium store; $k_f Y_i$ leaky Ca^{2+} transport from the calcium store to the cytosol; kZ_i Ca^{2+} transport from the cytosol to the extracellular environment. P (coupling parameter) is added to the original model to represent the degree of gap junctional connectivity between two adjacent cells. v_{2i} and v_{3i} are further described in the last two equations using rate constants, threshold constants and Hill coefficients (V_{M2}, K_2, n, V_{M3}, K_R, m, K_A, p). The model equations are numerically solved using the Euler's method in all the simulation analysis below.

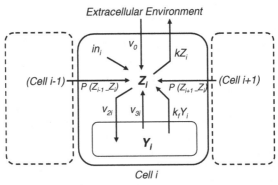

Fig. 6. Model of intercellular Ca^{2+} wave propagation

---

[2] The model predicts that Ca^{2+} oscillations can occur in the absence of IP_3 oscillations in [16].

5.1 Distance and Speed of Intercellular Ca²⁺ Waves

The distance and speed of traveling waves are first examined as a function of coupling parameter, P. In the following simulation, cell 0 is stimulated at time 0 for a short period of time, d. Therefore, $in_i = v_1 \cdot B$ for $i=0$ and for time 0 to d; and $in_i = 0$, otherwise. Two sets of parameter values prepared for simulation are as follows:

- **Parameter set1**: $v_0 = 1$ (uM/s), $v_1=7.3$ (uM/s), $B=0.01$, $k=3.5$ (1/s), $k_f=1$ (1/s), $V_{M2}=50$ (uM/s), $V_{M3} = 500$ (uM/s), $K_2=1$ (uM), $K_R=2$ (uM), $K_A=0.9$ (uM), $m=n=2$, $p=4$, $d=0.5$ (sec) and $n=50$. (P is varied.)
- **Parameter set2**: $V_{M2}=60$ (uM/s), $V_{M3} = 400$ (uM/s) and other values are the same as parameter set 1.

Figures 7 (a) and (b) plot the results obtained from parameter sets 1 and 2, respectively. In both cases, the coupling parameter (P) has a similar impact on the distance of waves. If P is very small (~0.15 for parameter set 1 and ~0.12 for parameter set 2), intercellular Ca²⁺ waves are either not generated or fail to travel (i.e., travel only a couple of cells). If P reaches a certain value (0.16 for parameter set 1 and 0.13 for parameter set 2), intercellular Ca²⁺ waves travel up to the last cell of the array (50th cell). As will be shown in Figure 9, the waves become regenerative in this case. If P is further increased (1.2 for parameter set 1 and 0.3 for parameter set 2),

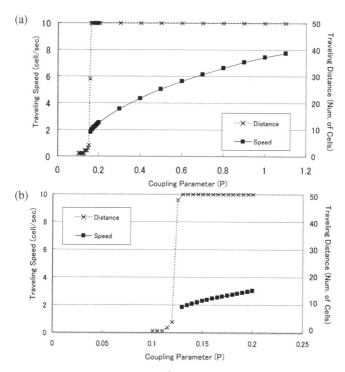

Fig. 7. Traveling distance and speed of Ca²⁺ waves from parameter sets 1 (a) and 2 (b)

intercellular Ca^{2+} waves are not generated at all. This is because the initial stimulating Ca^{2+} at cell 0 (i.e., in_0) diffuses so first to neighboring cells, that cell 0 and other cells fail to trigger Ca^{2+} release from the calcium store to increase cytosolic Ca^{2+} (thus, no first spike appearing at cell 0).

In both cases, the coupling parameter (P) has a similar impact on the speed of traveling waves. Larger P values increase the speed of intercellular Ca^{2+} waves traveling. Note in Figures 7 the traveling speed is measured only when waves became regenerative and propagated to the 50th cell.

Figures 8 and 9 depict two representative cases where intercellular Ca^{2+} waves fail to be regenerative (from parameter set1, $P=0.15$), and become regenerative (from parameter set1, $P=0.16$). In the case of propagation failure, the wave peak attenuates while traveling (Figure 8), while in the case of successful propagation the wave peak stays the same and propagates up to the last cell of the array (Figure 9). Note in Figure 9 that the peak of cell 0 is different from that of others. This is because cell 0 responds to in_i (the initial stimuli) to generate a spike, while cell i ($i{\neq}0$) responds to $P \cdot (Z_{i-1}-Z_i)$ (Ca^{2+} influx from cell i-1) to generate a spike.

As shown in the two sets of simulation results (from parameter sets 1 and 2), the speed and traveling distance of Ca^{2+} waves depend on not only the coupling parameter (P), but also cell properties (e.g.,. parameter values such as V_{M2} and V_{M3}).

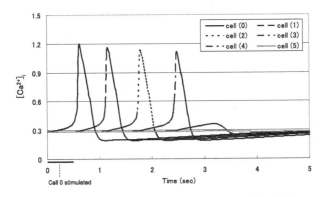

Fig. 8. Propagation failure (parameter set 1 and P=0.15)

5.2 Regenerative Intercellular Ca^{2+} Waves

We have next examined whether intercellular Ca^{2+} waves become regenerative or not under different parameter sets. For $k, V_{M2}, K_2, V_{M3}, K_R, K_A$, one parameter is examined at a time as a function of coupling parameter, P while other parameter values are the same as parameter set 1. Figures 10 show simulation results for each of the 6 parameters examined. In each figure, the parameter region enclosed with two lines (noted as the *regenerative* region) shows conditions under which intercellular Ca^{2+} waves become regenerative. As shown in the figures, the conditions are similarly found across all the 6 cases, and the case for k is explained here as a representative case.

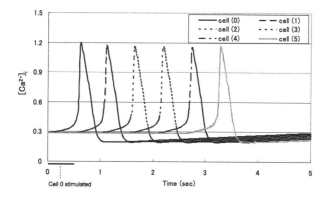

Fig. 9. Regenerative propagation (parameter set 1 and p=0.16)

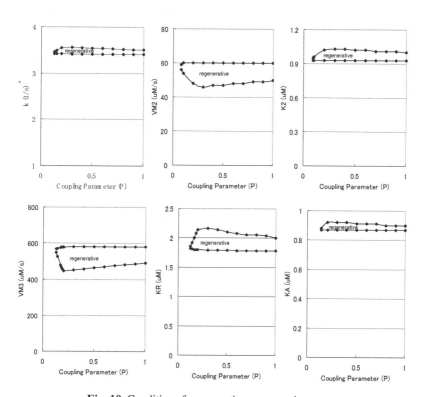

Fig. 10. Conditions for generating regenerative waves

When a small k values (k<3.42) is used, Z_i at rest state constantly generates a spike without stimulus and these regions are excluded from the regenerative region in the figure. This indicates that cells become *more excitable* for a smaller k value. For a small P value ($P < 0.13$), there is no parameter region that can generate regenerative waves. This is because a spike at cell 0 does not propagate to the next cell, or its spike peak attenuates while propagating cell-to-cell due to the low degree of gap junctional connectivity (see Figure 9). When the P value is between 0.13 and 0.3, the maximum k value of the regenerative region increases as the P value increases, growing the regenerative region with a higher degree of gap junctional connectivity. As the P value is further increased ($P>0.3$), however, the maximum k value of the regenerative region decreases. This is because a larger P value prohibits less excitable cells from generating the first spike that may propagate. As shown in the other figures, similar observations are made for the other five cases.

5.3   Repetitive Intercellular Ca^{2+} Waves

The last set of simulations has examined whether intercellular Ca^{2+} waves can be generated in response to repetitive stimuli. After cell 0 is stimulated at time 0 for a short period of time (d), cell 0 is again stimulated at $d + l$ for the same time period (d). Note here that l is the time interval between the end time of the initial stimulus and the start time of the 2nd stimulus. The parameter set 1 from subsection 6.1 is used for this simulation.

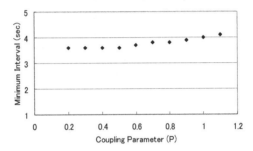

Fig. 11. Minimum time intervals to repetitively generate Ca^{2+} waves

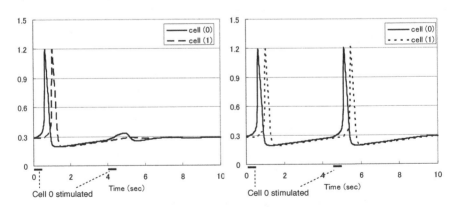

Fig. 12. Wave regeneration failure ($p=0.2$, $l=4.0$)

Fig. 13. Successful wave regeneration ($p=0.2$, $l=3.5$)

Figure 11 shows the minimal time interval (l) that is required to generate a regenerative Ca^{2+} wave in response to the 2nd stimulus, as a function of coupling parameter, P. The P values that do not produce regenerative waves are not examined nor plotted in this figure. Figures 12 and 13 show two selected results; figure 12 shows wave generation failure due to a short time interval between the two stimuli (l=3.5) (i.e., due to refractory dynamics of the Ca^{2+} oscillation model), and figure 13 successful wave generation with a sufficient time interval (l=4.0). The simulation result possibly implies the relationship between the coupling parameter and the data transmission rate; e.g., the number of spikes or *bits* that can be transmitted per time unit, showing that the data rate decreases as the coupling parameter increases.

5.4 Summary

The modeling study described in this section examined only limited aspects of physical layer issues (e.g. signal propagation), and a further study needs to be performed to obtain a deeper understanding of communication properties. Remaining work includes experimentally demonstrating the impact of gap junctional permeability predicted through this modeling study, and in addition, considering other models of calcium signaling proposed in the literature. In this study, a classical model of Ca^{2+} oscillation [16] is used, while other modes of calcium signaling may better represent calcium signaling observed in our experimental setup and are potentially useful to predict communication characteristics as well. In one model [20, 40] among other proposed models, IP_3 (i.e., Ca^{2+} mobilizing molecules) is generated in response to stimulating agents (chemical or mechanical stimulus), diffuses through gap junction channels from cell to cell, and stimulates Ca^{2+} stores (ER: Endoplasmic Reticulum) of each cell, which triggers Ca^{2+} release from the stores at each cell. As IP_3 diffuses cell-cell, intercellular Ca^{2+} waves propagate. In some other model [5], ATP is produced by stimulus, released into the extracellular environment, and reacts with membrane receptors of each cell. Each cell then produces Ca^{2+} mobilizing molecules (e.g., IP_3) and triggers Ca^{2+} release from Ca^{2+} stores. As ATP diffuses in the extracellular environment, intercellular Ca^{2+} waves propagate. In some other models, Ca^{2+} signals themselves propagate through gap junction channels and trigger Ca^{2+} release at each cell (calcium induced calcium release), thereby generating self-regenerative Ca^{2+} waves. Some other models further consider that Ca^{2+} may positively or negatively interact with IP_3 [43], and ATP may be regenerated [2]. All of these models are to be investigated in our further project together with biological experiments.

Remaining work also includes considering stochastic, spatial, and heterogeneous aspects in intracellular calcium wave propagation. Cellular processes generally involve stochastic processes and so does calcium signaling [17] (e.g., opening of Ca^{2+} release channels, random walk of Ca^{2+} signals and other molecules at low concentrations). The spatial aspect is an important factor influencing the nature of intercellular communication in tissue and organs [10, 25] and need to be considered. Cells are often highly heterogeneous in their shape, size, gap junctional connectivity and so forth. In contrast to all the above mentioned models, even simpler mathematical approximation such as [21] may be used to identify key communication related characteristics.

Future work also includes investigation into encoding processes [30, 39] where calcium spikes active target molecular machinery (e.g., calcium sensitive protein sensors), and application of information theory to quantify channel capacities of calcium signaling as studied for other biological communication media [12, 13, 37, 44].

6 Concluding Remarks

A line of research on molecular communication has started since the initial idea of molecular communication was presented in 2005 [18]. Research in molecular communication first approaches toward typical physical layer issues of computer communications (i.e., issues concerned with signal propagation). In addition to the molecular communication presented in this paper, other classes of signal propagation and molecular transport mechanisms are concurrently under developed (e.g., molecule transport by molecular motors [19]). Ongoing experimental work includes development of self-organizing molecular communication networks [32], engineering of various system components including information carriers [31] and sender/receiver cells [24]. To further advance this technology field, higher layer issues and system level issues will need to be addressed. Research questions of this kind include how molecular communication components can be integrated and interfaced to build larger scale complex systems and how stability and robustness of such integrated systems can be achieved.

It is also important to build theoretical foundations of molecular communication. The importance of information theory for small scale communication systems is addressed in [1]. Research efforts on information theory are being made for various classes of molecular communication (e.g., information theory for free diffusion media [3, 12, 13] and for signal transduction [28, 29]). These studies commonly investigate representation of information using molecules (i.e., molecular codes), maximal channel capacity of communication media, and effects of environmental noise on channel capacity. Theoretical foundations are indispensible not only to quantitatively understand how biological communications are performed, but also to engineer synthetic communication systems from biological systems.

Experimental and theoretical investigations of molecular communication all together may enable novel applications. Although not clearly demonstrated yet, molecular communication potentially impacts various technological domains including IT (e.g., unconventional computing and body sensor networks), health (e.g., nanomedicine and tissue engineering), environment (e.g., environmental monitoring) and military (e.g., biochemical sensors).

Acknowledgments

This research is supported by NiCT (National Institute of Information and Communications Technology, Japan), by the NSF through grants ANI-0083074, ANI-9903427 and ANI-0508506, by DARPA through grant MDA972-99-1-0007, by AFOSR through grant MURI F49620-00-1-0330, and by grants from the California MICRO and CoRe programs, Hitachi, Hitachi America, Hitachi CRL, Hitachi SDL, DENSO IT Laboratory, DENSO International America LA Laboratories, NTT Docomo and Novell.

References

[1] Alfano, G., Miorandi, D.: On information transmission among nanomachines. In: Proc. 1st International Conference on Nano-Networks and Workshops (2006)

[2] Anderson, C.M., Bergher, J.P., Swanson, R.A.: ATP-induced ATP release from astrocytes. J. Neurochem. 88, 246–256 (2004)

[3] Atakan, B., Akan, O.B.: An information theoretical approach for molecular communication. In: Proc. 2nd International Conference on Bio-Inspired Models of Network, Information, and Computing Systems (December 2007)

[4] Basu, S., Gerchman, Y., Collins, C.H., Arnold, F.H., Weiss, R.: A synthetic multicellular system for programmed pattern formation. Nature 434, 1130–1134 (2005)

[5] Bennett, M.V., Contreras, J.E., Bukauskas, F.F., Sáez, J.C.: New roles for astrocytes: gap junction hemichannels have something to communicate. Trends Neurosci. 26(11), 610–617 (2003)

[6] Berridge, M.J.: The AM and FM of calcium signaling. Nature 386, 759–780 (1997)

[7] Berridge, M.J., Bootman, M.D., Lipp, P.: Calcium - a life and death signal. Nature 395, 645–648 (1998)

[8] Berridge, M.J., Lipp, P., Bootman, M.D.: The versatility and universality of calcium signaling. Nature Reviews, Molecular Cell Biology 1, 11–21 (2000)

[9] Deitmer, J.W., Verkhratsky, A.J., Lohr, C.: Calcium signaling in glial cells. Cell Calcium 24(5-6), 405–416 (1998)

[10] Dokukina, I.V., Gracheva, M.E., Grachev, E.A., Gunton, J.D.: Role of mitochondria and network connectivity in intercellular calcium oscillations. Physica D (2007)

[11] Doktycz, M.J., Simpson, M.L.: Nano-enabled synthetic biology. Molecular Systems Biology 3, 125 (2007)

[12] Eckford, A.: Nanoscale communication with Brownian motion. In: Proc. 41st Annual Conference on Information Sciences and Systems (2007)

[13] Eckford, A.: Achievable information rates for molecular communication with distinct molecules. In: Proc. Workshop on Computing and Communications from Biological Systems: Theory and Applications (2007)

[14] Elfgang, C., Eckert, R., Lichtenberg-Frate, H., Butterweck, A., Traub, O., Klein, R.A., Hulser, D.F., Willeck, K.: Specific permeability and selective formation of gap junction channels in connexin-transfected HeLa cells. Journal of Cell Biology 129(3), 805–817 (1995)

[15] Freitas Jr., R.A.: Nanomedicine. Basic Capabilities, vol. I, Landes Bioscience (1999)

[16] Goldbetter, A., Dupont, G., Berridge, M.J.: Minimal model for signal-Induced Ca^{2+} oscillations and for Their frequency encoding through protein phosphorylation. Proceedings of the National Academy of Sciences 87, 1461–1465 (1990)

[17] Gracheva, M.E., Toral, R., Gunton, J.D.: Stochastic effects in intercellular calcium spiking in hepatocytes. Journal of Theoretical Biology 212, 111–125 (2001)

[18] Hiyama, S., Moritani, Y., Suda, T., Egashira, R., Enomoto, A., Moore, M., Nakano, T.: Molecular communication. In: Proc. 2005 NSTI Nanotechnology Conference (2005)

[19] Hiyama, S., Isogawa, Y., Suda, T., Moritani, Y., Suto, K.: A design of an autonomous molecule loading/ transporting/ unloading system using DNA hybridization and bimolecular linear motors in molecular communication. In: Proc. European Nano Systems (December 2005)

[20] Höfer, H., Venance, L., Giaume, C.: Control and plasticity of intercellular calcium waves in astrocytes: a modeling approach. Journal of Neuroscience 22(12), 4850–4859 (2002)

[21] Keener, J.P.: Propagation and its failure in coupled systems of discrete excitable cells. SIAM Journal on Applied Mathematics Volume 47(3), 556–572 (1987)

[22] Kinosita Jr., K., Yasuda, R., Noji, H., Adachi, K.: A rotary molecular motor that can work at near 100% efficiency. Philos. Trans. R. Soc. Lond. B Biol. Sci. 355, 473–489 (2000)

[23] Kitano, H.: Biological robustness. Nature Review Genetics 5, 826–837 (2004)

[24] Kikuchi, J., Sasaki, Y., Mukai, M., Moritani, Y., Hiyama, S., Suda, T.: Design of artificial cells for molecular communication. In: Proc. East Asian Biophysics Symposium & Annual Meeting of the Biophysical Society of Japan (EABS & BSJ) (November 2006)

[25] Kraus, M., Wolf, B., Wolf, B.: Crosstalk between cellular morphology and calcium oscillation patterns. Cell Calcium 19(6), 461–472 (1996)

[26] Krutovskikh, V.A., Piccoli, C., Yamasaki, H.: Gap junction intercellular communication propagates cell death in cancerous cells. Oncogene 21, 1989–1999 (2002)

[27] Laer, L., Kloppstech, M., Schofl, C., Sejnowski, T.J., Brabant, G., Prank, K.: Noise enhanced hormonal signal transduction through intracellular calcium oscillations. Biophysical Chemistry 91, 157–166 (2001)

[28] Liu, J.Q., Sawai, H.: A new channel coding algorithm based on photo-proteins and GTPases. In: 1st International Conference on Bio-Inspired Models of Network, Information, and Computing Systems (December 2006)

[29] Liu, J.Q., Nakano, T.: An information theoretic model of molecular communication based on cellular signalng. In: Proc. Workshop on Computing and Communications from Biological Systems: Theory and Applications (2007)

[30] Meyer, T., Stryer, L.: Calcium spiking. Annual Review of Biophysics and Biophysical Chemistry 20, 153–174 (1991)

[31] Moritani, Y., Hiyama, S., Suda, T.: Molecular communication among nanomachines using vesicles. In: Proc. 2006 NSTI Nanotechnology Conference (May 2006)

[32] Moore, M., Enomoto, A., Nakano, T., Egashira, R., Suda, T., Kayasuga, A., Kojima, H., Sakakibara, H., Oiwa, K.: A design of a molecular communication system for nanomachines using molecular motors. In: Proc. IEEE Conference on Pervasive Computing and Communications (2006)

[33] Moore, M., Enomoto, A., Suda, T., Nakano, T., Okaie, Y.: Molecular communication: new paradigm for communication among nano-scale biological machines. In: The Handbook of Computer Networks, vol. 3, pp. 1034–1054. John Wiley & Sons Inc, Chichester (2007)

[34] Nakano, T., Suda, T., Moore, M., Egashira, R., Enomoto, A., Arima, K.: Molecular communication for nanomachines using intercellular calcium signaling. In: Proc. 5th IEEE conference on nanotechnology (2005)

[35] Nakano, T., Suda, T., Koujin, T., Haraguchi, T., Hiraoka, Y.: Molecular communication through gap junction channels: system design, experiments and modeling. In: Proc. 2nd International Conference on Bio-Inspired Models of Network, Information, and Computing Systems (December 2007)

[36] Nakano, T., Hsu, Y.H., Tang, W.C., Suda, T., Lin, D., Koujin, T., Haraguchi, T., Hiraoka, Y.: Microplatform for intercellular communication. In: Proc. Third Annual IEEE International Conference on Nano/Micro Engineered and Molecular Systems (January 2008)

[37] Prank, K., Gabbiani, F., Brabant, G.: Coding efficiently and information rates in transmembrane signaling. BioSystems 55, 15–22 (2000)

[38] Peracchia, C.: Gap junctions: molecular basis of cell communication in health and diseases. Academic Press, London (2000)

[39] Salazar, C., Politi, A.Z., Höfer, T.: Decoding of calcium oscillations by phosphorylation cycles. Biophysical Journal 94, 1203–1215 (2008)

[40] Sneyd, J., Charles, A.C., Sanderson, M.J.: A model for the propagation of intercellular calcium waves. Cell Physiology 266(1), C293–C302 (1994)

[41] Sung, Y.J., Sung, Z., Ho, C.L., Lin, M.T., Wang, J.S., Yang, S.C., Chen, Y.J., Lin, C.H.: Intercellular calcium waves mediate preferential cell growth toward the wound edge in polarized hepatic cells. Exp. Cell Res. 287(2), 209–218 (2003)

[42] Tanenbaum, A.: Computer networks. Prentice-Hall, Englewood Cliffs (2003)

[43] Tsien, R.W., Tsien, R.Y.: Calcium channels, stores, and oscillations. Annu. Rev. Cell Biol. 6, 715–760 (1990)

[44] Thomas, P.J., Spencer, D.J., Hampton, S.K., Park, P., Zurkus, J.P.: The diffusion mediated biochemical signal relay channel. In: Proc. 17th Annual Conference on Neural Information Processing Systems (2003)

[45] Weiss, R., Basu, S., Hooshangi, S., Kalmbach, A., Karig, D., Mehreja, R., Netravali, I.: Genetic circuit building blocks for cellular computation, communications, and signal processing. Natural Computing 2, 47–84 (2003)

[46] Yang, G.-Z. (ed.): Body sensor networks. Springer, Heidelberg (2006)

[47] You, L., Ill, R.S.C., Weiss, R., Arnold, F.H.: Programmed population control by cell-cell communication and regulated killing. Nature 428, 868–871 (2004)

Clustering Time-Series Gene Expression Data with Unequal Time Intervals

Luis Rueda, Ataul Bari, and Alioune Ngom

School of Computer Science, University of Windsor
401 Sunset Avenue, Windsor, ON, N9B 3P4, Canada
{lrueda,bari1,angom}@uwindsor.ca

Abstract. Clustering gene expression data given in terms of time-series is a challenging problem that imposes its own particular constraints, namely exchanging two or more time points is not possible as it would deliver quite different results, and also it would lead to erroneous biological conclusions. We have focused on issues related to clustering gene expression temporal profiles, and devised a novel algorithm for clustering gene temporal expression profile microarray data. The proposed clustering method introduces the concept of profile alignment which is achieved by minimizing the area between two aligned profiles. The overall pattern of expression in the time-series context is accomplished by applying agglomerative clustering combined with profile alignment, and finding the optimal number of clusters by means of a variant of a clustering index, which can effectively decide upon the optimal number of clusters for a given dataset. The effectiveness of the proposed approach is demonstrated on two well-known datasets, yeast and serum, and corroborated with a set of pre-clustered yeast genes, which show a very high classification accuracy of the proposed method, though it is an unsupervised scheme.

Keywords: Microarrays, gene expression, time series data, clustering.

1 Introduction

Clustering genes based on the similarity of their temporal profile expressions has been studied for quite a few years. It is a problem that considers gene expression data given in terms of time series, and which is different from a general clustering problem, because exchanging two time points delivers quite different results, while it may not be biologically meaningful. This problem is important for many studies, such as determining genes that are functionally related or co-regulated [8]. Many unsupervised methods for gene clustering based on the similarity (or dissimilarity) of their microarray temporal profiles have been proposed in the past few years [2,5,6,9,18].

One of the methods for clustering microarray time-series data is based on a hidden phase model (similar to a hidden Markov model) to define the parameters of a mixture of normal distributions in a Bayesian-like manner, which are

C. Priami et al. (Eds.): Trans. on Comput. Syst. Biol. X, LNBI 5410, pp. 100–123, 2008.

estimated by using expectation maximization (EM) [2]. Clustering time-series data has also been studied using a Bayesian approach in [25], and a hidden Markov model (HMM) in [28]. A partitional clustering based on k-means and Euclidean distance has been studied in [31], and in [30], self-organizing maps (SOM) have been applied to visualize and interpret gene temporal expression profile patterns.

On the other hand, the method proposed in [5] requires computing the mean expression levels of some candidate profiles using some pre-identified, arbitrarily selected profiles. In [14], a method for clustering microarray time series data employing a jack-knife correlation coefficient with or without using the seeded candidate profiles is proposed. Specifying expression levels for the candidate profiles in advance for these correlation-based procedures requires estimating each candidate profile, which is made using a small sample of arbitrarily selected genes. This makes it vulnerable to the possibility of missing important genes, since the resulting clusters depend upon the initially chosen template genes. A regression-based approach to address the challenges for clustering short time-series expression datasets has been proposed in [9], which is suitable for analyzing single or multiple microarray temporal data. The study in [18] focused on analyzing gene temporal expression profiles datasets that are non-uniformly sampled and can have missing values. Gene temporal expression profiles are represented as continuous curves using statistical spline estimation. In [6], clustering of gene expression temporal profiles are done by identifying homogeneous clusters of genes. This method has focused on the *shapes of the curves* instead of the *absolute levels of expression*. Fuzzy Clustering gene temporal profiles has been the focus of [21], where the similarity measure for the co-expressed genes is computed based on the rate of change of the expression levels across time.

In [4], a regression-based approach is used to identify genes with different expression profiles across analytical groups in time-series experiments. The method is a two-step regression strategy where the experimental groups are identified by dummy variables. The procedure, in the first step, adjusts a global regression model with all the defined variables to identify differentially expressed genes. Statistically significant different profiles are found, in the next step, by applying a variable selection strategy that studies the differences between the groups.

Another method is to select and cluster genes using the ideas of order-restricted inference, where the estimation makes use of known inequalities among parameters [23]. First, potential candidate profiles of interest are defined and expressed in terms of inequalities between the expected gene expression levels at various time points. Second, for a given candidate profile, the estimated mean expression level of each gene is computed and the best fitting profile for a given gene is selected using the goodness-of-fit criterion and the bootstrap test procedure. Third, two genes expression profiles are assigned to the same cluster if they show similar profiles in terms of direction of the changes of expression ratios (e.g. up-up-up-down-down), regardless how big/small is the change.

In [1], a minimum-square-error profile alignment approach to cluster microarray time series data was proposed. The idea is to pairwisely align two temporal profiles in such a way that the sum of square errors between two aligned vectors is minimized. The alignment procedure, however, does not consider the length of the interval between two time points at which individual measurements are taken. In this paper, we propose a profile alignment approach to cluster temporal microarray data that minimizes the area between two aligned profiles [1]. The hierarchical clustering algorithm uses a variant of a well-known clustering validity index that optimizes the number of clusters [20]. The profile alignment that we propose in this paper is different from that of [1] in the sense that: (i) the approach proposed in this paper considers unequal time intervals, which is usually the case in microarray time-series experiments, and (ii) the alignment is performed by minimizing the error between two continuous functions and not the "knot" points. Experiments on serum data and on pre-clustered yeast data show the effectiveness of the proposed method.

2   Area-Based Profile Alignment

Clustering time-series expression data with unequal time intervals is a very special problem, as measurements are not necessarily taken at equally spaced time points. Taking into account the length of the interval is accomplished by means of analyzing the area between two expression profiles, joined by the corresponding measurements at subsequent time points. This is equivalent to considering the sum or average of square errors between the infinite points in the two lines. This analysis can be easily achieved by computing the underlying integral, which is analytically resolved in advance, subsequently avoiding expensive computations during the clustering process.

Consider a dataset $\mathcal{D} = \{\mathbf{x}_1, \mathbf{x}_2, .., \mathbf{x}_n\}$, where $\mathbf{x}_i = [x_{i_1}, x_{i_2}, ..., x_{i_m}]^t$ is an m-dimensional feature vector that represents the expression ratio of gene i at m different time points, $\mathbf{t} = [t_1, t_2, ..., t_m]^t$. The aim is to partition $\mathcal{D}$ into k disjoint subsets $\mathcal{D}_1, \mathcal{D}_2, ..., \mathcal{D}_k$, where $\mathcal{D} = \mathcal{D}_1 \cup \mathcal{D}_2 \cup ... \cup \mathcal{D}_k$, and $\mathcal{D}_i \cap \mathcal{D}_j = \emptyset$, for $\forall i, j, i \neq j$, in such a way that a similarity (dissimilarity) cost function $\phi : \{0, 1\}^{n \times k} \to \Re$ is maximized (minimized).

Our profile alignment approach takes into consideration the length of the intervals, which could be unequal, between the time points at which the measurements are taken in an experiment. The idea of this scheme works as follows.

Let $\mathbf{t} = [t_1, t_2, ..., t_m]^t$ be the vector representing the time points at which the measurements are taken, and let $\mathbf{x} = [x_1, x_2, ..., x_m]^t$, and $\mathbf{y} = [y_1, y_2, ..., y_m]^t$ be two expression profiles, whose expression ratios were measured at time points given in $\mathbf{t}$, which are to be aligned. The aim is to find a scalar a that minimizes the total area between the two profiles, e.g., between the lines that join the expression ratios. To do this, first, $\mathbf{x}$ is aligned in a transformed space, and a new vector, $\mathbf{x}'$, is obtained as follows:

$$\mathbf{x}' = [x_1', x_2', ..., x_m']^t \leftarrow \mathbf{x} - x_1 . \qquad (1)$$

---

[1] A preliminary version of this work appeared in [27].

Now, let the straight line that joins points (t_{i-1}, x'_{i-1}) and (t_i, x'_i) be denoted by $x'_{i-1} + \frac{(x'_i - x'_{i-1})}{t_i - t_{i-1}} u$, and for points (t_{i-1}, y_{i-1}) and (t_i, y_i) be denoted by $y_{i-1} - a + \frac{(y_i - y_{i-1})}{t_i - t_{i-1}} u$. Since the aim is to minimize the area between $\mathbf{x}$ and $\mathbf{y}$ (aligned), for all $t_1, t_2, ..., t_m$, it is necessary to find a value of a that minimizes the sum of square errors between each pair of points in the continuous times, equivalent to the following sum of integrals:

$$f(a) = \sum_{i=2}^{m} \int_{t_{i-1}}^{t_i} \left[x'_{i-1} + a - y_{i-1} + \frac{(x'_i - x'_{i-1}) - (y_i - y_{i-1})}{t_i - t_{i-1}} u \right]^2 du, \qquad (2)$$

by means of the first and second order conditions. Solving the integral and isolating a results in:

$$a = -\frac{\sum_{i=2}^{m} \left[\left(x'_{i-1} - y_{i-1} \right) \left(t_i - t_{i-1} \right) + \frac{(x'_i - x'_{i-1}) - (y_i - y_{i-1})}{t_i - t_{i-1}} \frac{(t_i - t_{i-1})^2}{2} \right]}{\sum_{i=2}^{m} (t_i - t_{i-1})}. \qquad (3)$$

Then, a new vector, $\mathbf{y}'$, is computed as follows:

$$\mathbf{y}' = \mathbf{y} - a. \qquad (4)$$

Letting $f_i = \frac{(x'_i - x'_{i-1}) - (y'_i - y'_{i-1})}{t_i - t_{i-1}}$, and $\tau = t_i - t_{i-1}$, and by computing the integral distance between the two new vectors $\mathbf{x}'$ and $\mathbf{y}'$, $d(\mathbf{x}', \mathbf{y}')$, results in:

$$d\left(\mathbf{x}', \mathbf{y}'\right) = \sum_{i=2}^{m} \left[\left(x'_{i-1} - y'_{i-1} \right)^2 u + \left(x'_{i-1} - y'_{i-1} \right) f_i u^2 + \frac{f_i^2 u^3}{3} \Big|_{t_{i-1}}^{t_i} \right]$$

$$= \sum_{i=2}^{m} \left(x'_{i-1} - y'_{i-1} \right)^2 \tau + \left(x'_{i-1} - y'_{i-1} \right) f_i \tau^2 + f_i^2 \frac{\tau^3}{3}. \qquad (5)$$

It is not difficult to show that the second order condition is satisfied: $\frac{\partial^2 f}{\partial^2 a} > 0$. Using (3) to obtain the value of a, two profile vectors are aligned first, and then a distance function is computed in the usual manner. It can be shown that the alignment, used in conjunction with any metric d, is also a metric [1]. It may be noted that once the alignment is applied, any metric d can be used.

The idea is described in Fig. 1. In Fig. 1(a) two vectors are shown before alignment. Fig. 1(b) shows the "aligned" vectors such that the area between the profiles is minimized, i.e. they were aligned in such a way that the total area covered by the triangle $\{u, v, z\}$ and the polygon $\{z, w, q, e, r, k, h, g, s\}$ is minimized.

The aim of clustering gene expression temporal data is to group together profiles with similar patterns. Deciding upon the similarity often involves pairwise distance measures of co-expressions. Conventional distance measures include correlation, rank correlation, Euclidean distance, angle between vectors of observations, among others. However, clustering algorithms that apply a conventional

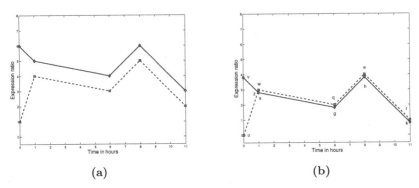

Fig. 1. (a) Two unaligned profiles. (b) The two "aligned" profiles obtained after applying (1) - (4) such that the area between each pair of lines is minimized.

distance function (e.g. the Euclidian distance, correlation coefficient) on a pair of unaligned profiles, can be too general to reflect the temporal information embedded in the underlying expression profiles. Some alignment techniques can be used to resolve the issue effectively prior to applying the distance function.

Typical examples for the profile alignment are depicted in Fig. 2 with different cases of temporal profiles for three pairs of genes. Fig. 2(a), (b) and (c) show three pairs of genes prior to alignment. Using the Pearson correlation distance, genes in (b) are more likely to be clustered together, since they produce the largest value for the correlation coefficient among all three pairs of genes, which yields 0.9053. If the prime interest is to cluster genes according to the variation of their expression level at different time points, then, genes from (c) would be better candidates to be clustered together than the genes in (a) and (b). However, the value of the correlation coefficient between the pairs of genes in (c) is the minimum (0.8039) among all three pairs of genes. Fig. 2(d), (e) and (f) show the pairs of genes after aligning the expression profiles from (a), (b) and (c), respectively. A careful visual inspection shows that the expression profiles in (f) are closer to each other compared to the expression profiles from the other two figures, (d) and (e).

2.1   Clustering Method for Profile Alignment

Many clustering approaches have been proposed so far, including the methods proposed in [5,8,17,29]. Each of these methods has its own pros and cons depending upon handling noise levels in the measurements and the properties of the particular dataset being clustered, and hence, none of them can be regarded as *the best* method.

We use agglomerative clustering, where the decision rule is based on the *furthest-neighbor* distance between two clusters [7], which is computed using the area-based distance as in (5). This distance involves the alignment of each pair of profiles before applying a conventional distance function. The procedure is formalized in Algorithm *Agglomerative-Clustering*, which is the general

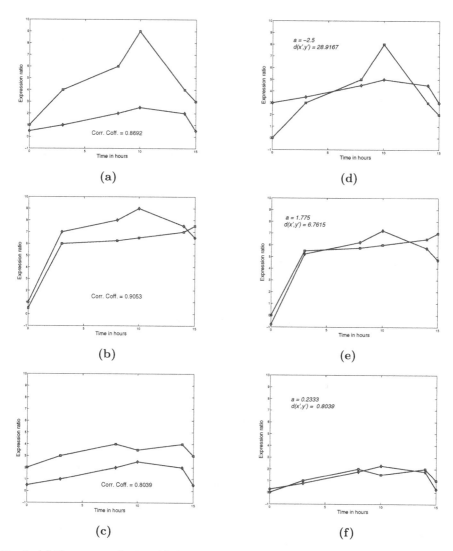

Fig. 2. (a) Two genes that are likely to be clustered as in [23], although the difference among them in terms of rate of expression ratio changes between different time points is large; (b) Two genes with different profiles that are likely to be clustered together by correlation-based methods; (c) Two genes with similar profiles in terms of rate of expression ratio changes between different time points that may not be clustered together by the method proposed in [23] and the correlation-based methods; (d), (e) and (f) depict the result after aligning the two genes from (a), (b) and (c), respectively

hierarchical clustering algorithm modified to compute the best number of clusters. The algorithm receives two parameters as input, a complete microarray temporal dataset, $\mathcal{D}$, and the desired number of clusters, k, and returns the dataset after partitioning it into k clusters.

Algorithm 1. Agglomerative-Clustering

Input: The dataset, $\mathcal{D} = \{\mathbf{x}_1, \mathbf{x}_2, ..., \mathbf{x}_n\}$, and k, the desired number of clusters.
Output: k disjoint subsets $\mathcal{D}_1, \mathcal{D}_2, ..., \mathcal{D}_k$.
begin
    Create n clusters, $\mathcal{D}_1, \mathcal{D}_2, ..., \mathcal{D}_n$, where $\mathcal{D}_i = \{\mathbf{x}_i\}$
    $\mathcal{D}_{currentClustersSet} \longleftarrow \{\mathcal{D}_1, \mathcal{D}_2, ..., \mathcal{D}_n\}$
    for $q \longleftarrow n$ **down to** k **do**
        For each pair of cluster $(\mathcal{D}_i, \mathcal{D}_j)$, find furthest neighbor $(\mathbf{x}_i, \mathbf{x}_j)$, using the profile alignment and then computing the distance between them.
        Select $(\mathcal{D}_j, \mathcal{D}_l)$ as the pair of clusters with the closest furthest neighbors.
        $\mathcal{D}_{mergedClusters} \longleftarrow \{\mathcal{D}_j \cup \mathcal{D}_l\}$
        $\mathcal{D}_{currentClustersSet} \longleftarrow \{\mathcal{D}_{currentClustersSet} \cup \mathcal{D}_{mergedClusters}\} \setminus \{\mathcal{D}_j, \mathcal{D}_l\}$
    end
end
return $\mathcal{D}_{currentClustersSet}$

2.2 Optimizing the Number of Clusters

Finding the optimal number of clusters is a well known open problem in clustering. For this, it is desirable to validate the number of clusters that is the best for a given dataset using a validity index. For the validity purpose of the clustering, a variant of the $\mathcal{I}$-index [20] is proposed. The traditional $\mathcal{I}$-index is defined as follows:

$$\mathcal{I}(k) = \left(\frac{1}{k} \times \frac{E_1}{E_k} \times \mathcal{D}_k \right)^p, \tag{6}$$

where, $E_k = \sum_{i=1}^{k} \sum_{j=1}^{n} u_{ij} \|\mathbf{x}_j - \boldsymbol{\mu}_i\|$, $\mathcal{D}_k = max_{i,j=1}^{k} \|\boldsymbol{\mu}_i - \boldsymbol{\mu}_j\|$, n is the total number of samples in the dataset, $\{u_{ij}\}_{k \times n}$ is the partition (or membership) matrix for the data, $\boldsymbol{\mu}_i$ is the center of cluster $\mathcal{D}_i$ and k is the number of clusters. The contrast between different cluster configurations is controlled by the exponent p (usually, $p = 2$). The best number of clusters k^* is chosen to maximize $\mathcal{I}(k)$.

Although this index has been found to work well in many cases, a superficial analysis of $\mathcal{I}(k)$ suggests that using the value of $p = 2$ for the factor $\frac{1}{k}$ penalizes large number of clusters, and hence, contradicts the aim of finding differentially expressed genes, which in many cases, needs to form clusters containing only one or two genes [1,27]. Therefore, we propose the following variant of the $\mathcal{I}$-index for Profile Alignment and Agglomerative Clustering (PAAC) methods [1]:

$$\mathcal{I}_{PAAC}(k) = \left(\frac{1}{k} \right)^q \times \left(\frac{E_1}{E_k} \times \mathcal{D}_k \right)^p, \tag{7}$$

where $q < p$. The partition matrix $\{u_{ij}\}$ is defined as a membership function such that $u_{ij} = 1$, if $\mathbf{x}_j$ belongs to cluster $\mathcal{D}_i$, and zero otherwise. Again, the best number of clusters is the value of k that maximizes $\mathcal{I}_{PAAC}(k)$.

The implementation of index $\mathcal{I}_{PAAC}(k)$ is not straightforward. It includes the mean and scatter for each cluster, and these two are meant to include the profile alignment concept. For the mean, algorithm *Cluster-Mean* is used. The algorithm arbitrarily selects a gene from the cluster as the initial mean, then iteratively (i) aligns the next gene profile to the current mean, (ii) computes a new mean by taking the average of the two aligned profile. The process continues until all genes, from the cluster, are considered. Note that computing the mean in this way has the particularity that changing the order of the genes produces different means. Also, the mean obtained does not represent the *actual* mean; it is just an estimate. Another way of computing the mean would be, for example, by aligning *all* profiles in a cluster and then, compute the mean. However, this strategy for alignment includes multiple alignment, which remains an open problem.

Algorithm 2. Cluster-Mean

Input: A cluster $\mathcal{D}_i$ with n_i samples $\mathcal{D}_i = [\mathbf{x}_{i_1}, \mathbf{x}_{i_2}, \ldots, \mathbf{x}_{i_{n_i}}]$.
Output: The mean of cluster $\mathcal{D}_i$, $\boldsymbol{\mu}_i$.
begin
  $\boldsymbol{\mu}_i \longleftarrow \mathbf{x}_{i_1}$
  for $j \longleftarrow 2$ **to** n_i **do**
    $[\mathbf{y}_1, \mathbf{y}_2] \longleftarrow d_{PAAC}(\boldsymbol{\mu}_i, \mathbf{x}_{i_j})$
    $\boldsymbol{\mu}_i \longleftarrow \frac{1}{2}(\mathbf{y}_1 + \mathbf{y}_2)$
  end
end
return $\boldsymbol{\mu}_i$

Once the *Cluster-Mean* is defined, it is used to compute the *scatter* of a cluster, using the algorithm *Within-Cluster-Scatter*. The algorithm takes a cluster of gene expression profile and its mean profile as inputs, and computes the sum of the distances between each gene expression profile and the cluster mean. Each expression profile and the mean profile are aligned before a distance function is applied.

Algorithm 3. Within-Cluster-Scatter

Input: A cluster $\mathcal{D}_i$ with n_i samples, $\mathcal{D}_i = [\mathbf{x}_{i_1}, \mathbf{x}_{i_2}, \ldots, \mathbf{x}_{i_{n_i}}]$, and its mean, $\boldsymbol{\mu}_i$.
Output: The sum of the distances of each gene from the cluster mean, E_i.
begin
  $E_i \longleftarrow \mathbf{0}$
  for $j \longleftarrow 1$ **to** n_i **do**
    $E_i \longleftarrow E_i + d_{PAAC}(\boldsymbol{\mu}_i, \mathbf{x}_{i_j})$
  end
end
return E_i

---

3 Experimental Results

In this section, we show the performance of PAAC on two well-known datasets, the serum [17] and the yeast [5] datasets. The results are compared with that of agglomerative clustering when used with the Pearson and the Spearman correlations to measure the similarity between two profiles. We also investigate the biological significance of PAAC on the dataset of [3].

The best number of clusters, for each dataset, is computed using the index I_{PAAC} defined in (7). It is well-known that, in general, the best number of clusters for any real-life detaset is usually less than or equal to $\sqrt{n}$, where n is the number of samples in the dataset [20]. However, in time-series gene expression data, some useful clusters contain one or two differentially expressed genes. Thus, a range for potential numbers of clusters, which includes the values of k that lie between $\left\lceil \sqrt{\frac{1}{2}n} \right\rceil$ and $\left\lfloor \sqrt{\frac{3}{2}n} \right\rfloor$, is considered.

3.1   The Serum Dataset

The serum dataset, obtained from the experimental data used in [17], contains data on the transcriptional response of cell cycle-synchronized human fibroblasts to serum. These experiments have measured the expression levels of 8,613 human genes after a serum stimulation at twelve different time points, at 0 hr., 15 min., 30 min., 1 hr., 2 hrs., 3 hrs., 4 hrs., 8 hrs., 16 hrs., 20 hrs. and 24 hrs. From these 8,613 gene profiles, 517 profiles were separately analyzed, as their expression ratio has changed substantially at two or more time points. The experiments and analysis have focused on this dataset[2], which is the same group of 517 genes used in [17].

The best number of clusters is obtained using I_{PAAC}, where the distance is computed by using $d_{MSE}(.,.)$ [1]. For the serum dataset, the range for the number of clusters was set to $k = 16$ to 27. The values of the I_{PAAC} index were computed for each of these values of k. We keep the value of p fixed at 2 in all computations of the index. Using I_{PAAC} for each k, we tried values of q from 0.3 to 1.0 and found that the value of I_{PAAC} reaches a maximum when $q = 0.7$ and $k = 21$. Therefore, we took $k = 21$ as the best number of cluster.

To compare the results, the serum dataset was clustered into 21 clusters, using the Pearson correlation distance within the same hierarchical agglomerative clustering method, and the PAAC method. The 21 clusters of the serum dataset, using the PAAC method corresponding to $k = 21$, are plotted in Fig. 3. The x-axis in each plot represents the time in hours and the y-axis represents the expression ratio. Each plot represents a cluster. The resulting 21 clusters found using the Pearson correlation are shown in Fig. 4.

The comparison among the plots for PAAC and Pearson (Fig. 4) reveal the effectiveness of our PAAC method to discover new profiles. For example, PAAC left clusters 1 to 5 containing a single gene each (IDs 328692, 470934, 361247, 147050 and 310406, respectively). The Pearson correlation method, however, placed these genes in clusters 19, 2, 2, 11 and 19, respectively. By visual inspection of these Pearson clusters, it can be observed that these genes are *differentially expressed* and should be left alone in separate clusters, which is clearly done by PAAC. Also, PAAC produced four clusters containing only two profiles each, clusters 9 (IDs 356635 and 429460), 11 (IDs 26474 and 254436), 13 (IDs 280768 and 416842) and 16 (IDs 130476 and 130482). The Pearson correlation

---

[2] Expression data for this subset were obtained from the website: http://genome-www.stanford.edu/serum/.

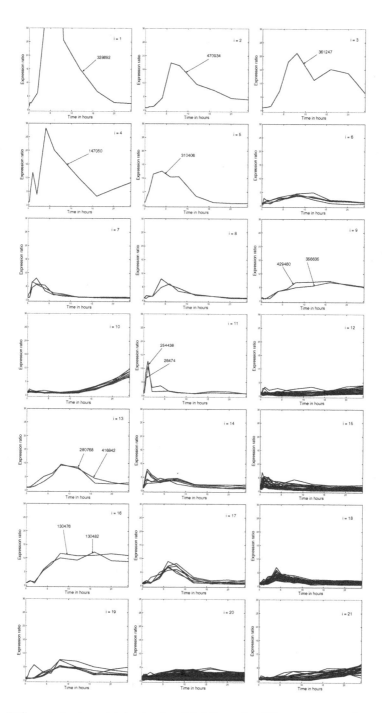

Fig. 3. Different clusters obtained using PAAC on the 517 gene temporal expression profiles, where $k = 21$

method clustered these genes as follows: 356635 and 429460 in cluster 16, 26474 and 254436 in cluster 21, 280768 and 416842 in cluster 2 and 130476 and 130482 in cluster 16. Although the Pearson correlation method placed each pair of genes in the same cluster, it also placed some other genes with them. By looking at the plots of the profiles of the clusters produced by the Pearson correlation method and comparing them to the plots of the clusters of the corresponding genes produced by PAAC, it is clear that these pairs of genes are differentially expressed.

We also compared PAAC with the results obtained using the Spearman correlation as a distance measure (see Fig. 5). It can be observed that the Spearman correlation, though non-linear, is not able to identify and separate the differentially expressed genes properly. This situation is clearly observed in clusters 5, 6, 12, 18, and 21.

3.2   The Yeast Dataset

The yeast dataset contains the changes in gene expression ratios obtained during sporulation of 6,118 yeast genes[3] [5]. The expression levels in the dataset were measured at seven different time points, at 0 hr., 0.5 hr., 2 hrs., 5 hrs., 7 hrs., 9 hrs. and 11.5 hrs.

For the yeast dataset, the values of the $\mathcal{I}_{PAAC}$ index were computed for $k = 55$ to 95, while keeping the range of values for q the same as before (from 0.3 to 1.0). Again, the value of $q = 0.7$ was selected, for which the value of the index reaches to a maximum level when $k = 60$. Therefore, $k = 60$ was taken as the optimal number of clusters and the profiles are plotted, as clustered using the PAAC method, and the Pearson correlation.

The plots of the clusters, obtained using the PAAC method are shown in Fig. 6 and 7. Each figure shows 30 out of 60 clusters. The plots of the resulting 60 clusters using the Pearson correlation distance are shown in Fig. 8 and 9. Again, each figure shows the plots of 30 out of 60 clusters. Form the figures, it can be observed that although PAAC produced a number of clusters containing a single gene each, these genes are very differently expressed as compared to the other genes in the dataset. Therefore, we argue that these genes are better to be placed in individual clusters. For example, genes YER150w, YGL055W, YGR142W, YKL152C, YKL153W, YLR194C, YLR377C, YMR316W and YPR074C were placed in clusters with a single sample each by PAAC ($i = 34, 37, 40, 51, 52, 53, 56, 58,$ and 60, respectively). It is clear that all these expression profiles are completely different and hence they should be placed alone in a single cluster. The Pearson correlation method, on the other hand, placed these genes in clusters 45, 4, 1, 28, 15, 6, 38, 37 and 6, respectively ($i = 45, 4, 1, 28, 15, 6, 38, 37$ and 6 in Fig. 8 and 9). However, simply looking at these figures of the Pearson correlation distance, in terms of expression ratio changes, each of these genes has very different temporal profiles and can be easily distinguished from the rest of the genes placed in

---

[3] Expression   data   for   this   dataset   were   obtained   from   the   website:
   http://cmgm.stanford.edu/pbrown/sporulation/additional/.

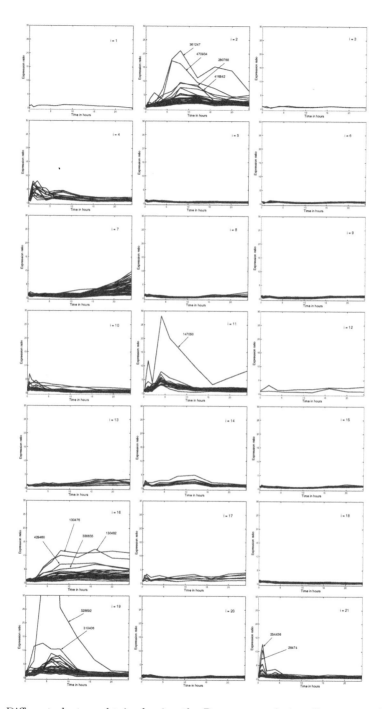

Fig. 4. Different clusters obtained using the Pearson correlation distance on the 517 gene temporal expression profiles, where $k = 21$

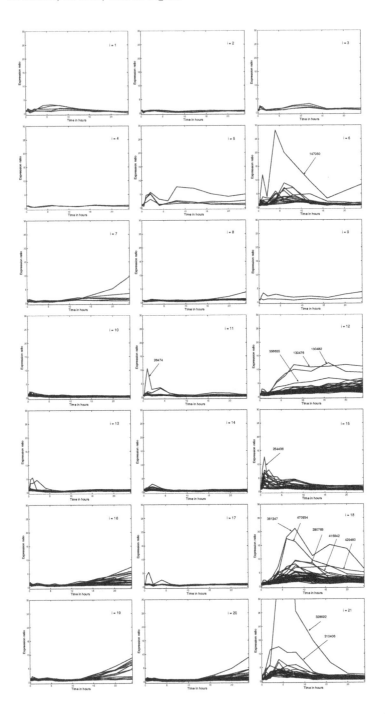

Fig. 5. Different clusters obtained using the Spearman correlation distance on the 517 gene temporal expression profiles, where $k = 21$

the same cluster by the Pearson correlation method. Therefore, PAAC produced clusters with better intra-cluster profile similarity.

3.3 Biological Analysis

In order to provide a biological significance of the results obtained, PAAC was applied to a dataset containing the changes in gene expression during the cell cycle of the budding yeast S. cerevisiae[4] [3]. The expression levels in the dataset were measured at seventeen different time points, from 0 min. to 160 min. with an interval of 10 min. The experiment monitored 6,220 transcripts for cell cycle-dependent periodicity, and 221 functionally characterized genes with periodic fluctuation were listed in Table 1 of [3].

The I_{PAAC} index was computed on the this dataset by keeping the parameter the same as those for serum dataset and for q=0.7, the best number of clusters $k = 28$ was obtained. The clusters obtained using PAAC are shown in Fig. 10. The same dataset was also clustered using the Pearson correlation coefficient, and the plots are shown in Fig. 11. A comparison of the two figures shows that PAAC separates the genes by profiles in a wise manner, while the Pearson correlation method is not able to capture all variations in the time series. This can be observed in various clusters that include genes with quite different profiles.

To show the biological significance of the results, the 221 genes are listed in Table 1, where, for each gene, the cluster number that PAAC assigns to a gene and the actual phase (class) of that same gene, as categorized in Table 1 of [3], is shown. An objective measure for comparing the two clusterings was taken by computing the overall classification accuracy, as the number of genes that PAAC *correctly* assigned to one of the phases. The *correct* class (phase) is the one that PAAC assigns the largest number of genes. This criterion is reasonable, as PAAC is an unsupervised learning approach that does not know the classes beforehand, and hence the aim is to "discover" new classes. The overall classification accuracy was computed as the average of the individual accuracies for each cluster, resulting in 83.47%. This accuracy is very high considering the fact that PAAC is an *unsupervised* classification algorithm.

---

[4] Expression data for this dataset were obtained from the website: http://genome-www.stanford.edu/cellcycle/data/rawdata/individual.html.

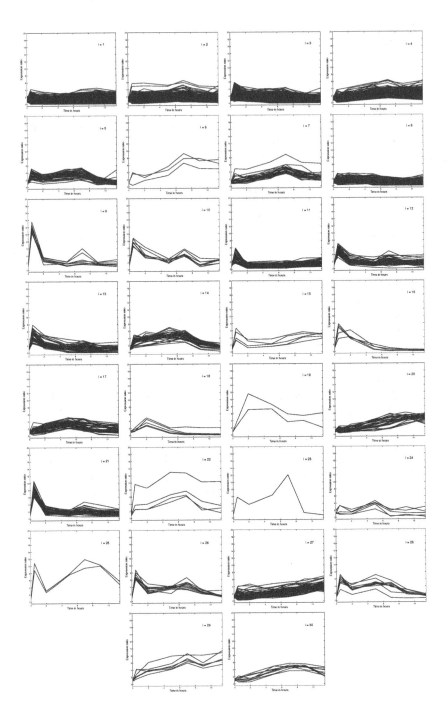

Fig. 6. Clusters $i = 1$ to 30, obtained using the PAAC on the 6,118 gene temporal expression profiles of the yeast dataset

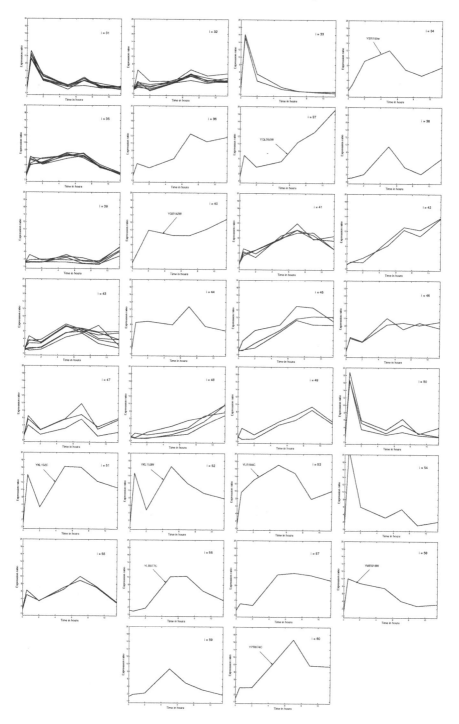

Fig. 7. Clusters $i = 31$ to 60, obtained using the PAAC on the 6,118 gene temporal expression profiles of the yeast dataset

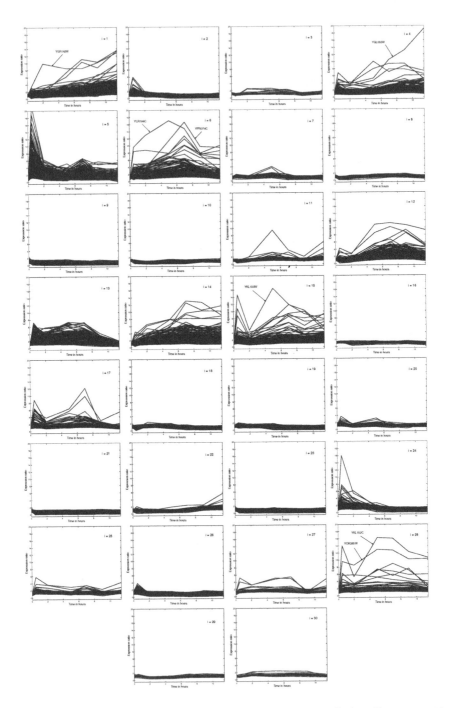

Fig. 8. Clusters $i = 1$ to 30, obtained using the Pearson correlation distance on the 6,118 gene temporal expression profiles of the yeast dataset

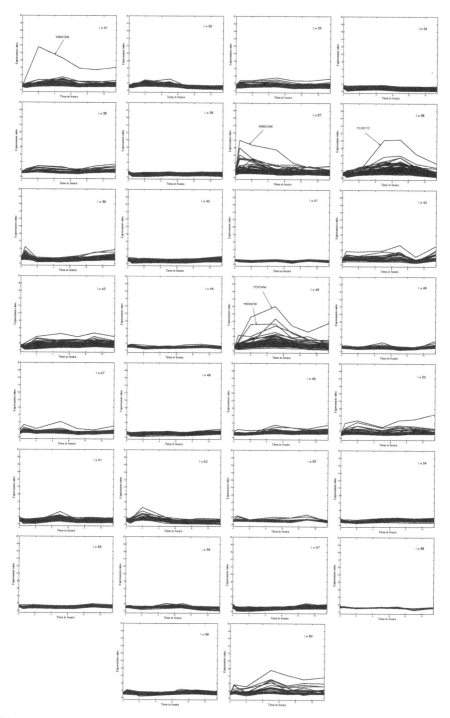

Fig. 9. Clusters $i = 31$ to 60, obtained using the Pearson correlation distance on the 6,118 gene temporal expression profiles of the yeast dataset

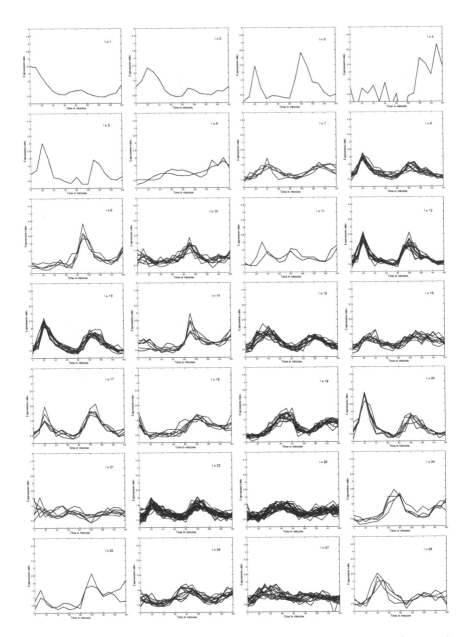

Fig. 10. Different clusters obtained using PAAC on the 221 gene temporal expression profiles from Table 1 of [3], where $k = 28$

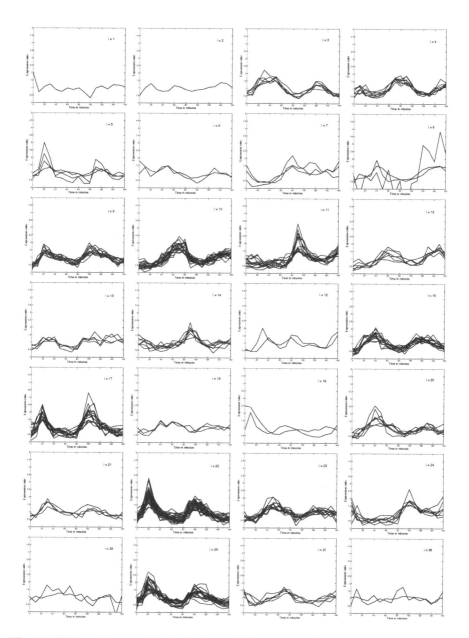

Fig. 11. Different clusters obtained using the Pearson correlation coefficient on the 221 gene temporal expression profiles from Table 1 of [3], where $k = 28$

Table 1. Genes from Table 1 of [13], clustered using PAAC, where $k = 28$

PAAC	Gene	Phase	PAAC	Gene	Phase	PAAC	Gene	Phase
1	YBR067c/TIP1	Early G_1	15	YEL061c/CIN8	S	22	YLL002w/	Late G_1
2	YGL055W/OLE1	Early G_1	15	YGR140W/CBF2	S	22	YLR457C/NBP1	Late G_1
3	YDL227C/HO	Late G_1	15	YHR172W/	S	22	YPL057C/SUR1	Late G_1
4	YAL001C/TFC3	S	15	YLR045c/STU2	S	22	YPL124W/NIP29	Late G_1
5	YJL115W/ASF1	S	15	YNL126W/	S	22	YBR275c/RIF1	S
6	YDL198C/SHM1	G_2	15	YPR141C/KAR3	S	22	YCR065w/HCM1	S
6	YDR146c/SWI5	M	15	YJR006W/HUS2	S	22	YDL197C/ASF2	S
7	YKL049C/CSE4	G_2	15	YER001w/MNN1	S	22	YKL127W/PGM1	S
7	YKL048C/ELM1	G_2	15	YER003c/PMI40	S	22	YDL095W/PMT1	S
7	YER069w/"ARG5,6"	G_2	16	YNL225C/	Late G_1	23	YLR210W/CLB4	S
7	YJR112W/NNF1	G_2	16	YDR224c/HTB1	Late G_1	23	YDR150w/NUM1	M
7	YPR167C/MET16	M	16	YDR225w/HTA1	Late G_1	23	YMR198W/CIK1	S
8	YFL008W/SMC1	Late G_1	16	YGL200C/EMP24	Late G_1	23	YBL052c/	S
8	YMR078C/CHL12	Late G_1	16	YNL272C/SEC2	Late G_1	23	YIL126W/STH1	S
8	YPL209C/IPL1	Late G_1	16	YDR488c/PAC11	S	23	YCR035c/	S
8	YGR152C/RSR1	Late G_1	16	YBL002w/HTB2	S	23	YIL050W/	G_2
8	YIL159W/	Late G_1	16	YBL003c/HTA2	S	23	YBL097w/	G_2
8	YNL233W/	Late G_1	16	YKL067W/YNK1	S	23	YJL099W/CHS6	G_2
8	YKL045W/PRI2	Late G_1	16	YER118c/SSU81	S	23	YJR076C/CDC11	G_2
8	YNL312W/RFA2	Late G_1	17	YJL074C/SMC3	Late G_1	23	YCR084c/TUP1	G_2
8	YML061C/PIF1	Late G_1	17	YGR041W/	Late G_1	23	YGL255W/ZRT1	G_2
8	YLR233C/EST1	Late G_1	17	YNL082W/PMS1	Late G_1	23	YJL137c/GLG2	G_2
8	YOR026W/BUB3	S	17	YOL090W/MSH2	Late G_1	23	YCR073c/	G_2
9	YDL179w/	Early G_1	17	YHR153c/SPO16	Late G_1	23	YDR389w/SAC7	G_2
9	YLR079w/SIC1	Early G_1	18	YCR005c/CIT2	Early G_1	23	YKL068W/NUP100	G_2
9	YJL157C/FAR1	Early G_1	18	YCL040w/GLK1	Early G_1	23	YGR092W/DBF2	M
9	YKL185W/ASH1	Early G_1	18	YLR258W/GSY2	Early G_1	23	YOR058C/ASE1	M
10	YJL194W/CDC6	Early G_1	18	YNL173C/	Late G_1	23	YPL242C/	M
10	YLR274W/CDC46	Early G_1	18	YOR317W/FAA1	Late G_1	23	YCL037c/SRO9	M
10	YPR019W/CDC54	Early G_1	18	YNL073W/MSK1	S	23	YKL130C/	M
10	YHR005c/GPA1	Early G_1	19	YIL106W/MOB1	G_2	23	YNL053W/MSG5	M
10	YGR183C/QCR9	Early G_1	19	YCL014w/BUD3	G_2	23	YIL162W/SUC2	M
10	YLR273C/PIG1	Early G_1	19	YGR108W/CLB1	M	23	YDL048c/STP4	M
10	YLL040c/	Early G_1	19	YPR119W/	M	23	YHR152W/SPO12	M
10	YHR038W/	Late G_1	19	YBR138c/	M	23	YKL129C/MYO3	M
10	YAL040C/CLN3	M	19	YHR023w/MYO1	M	24	YPL058C/PDR12	Early G_1
11	YDR277c/MTH1	S	19	YOL069W/NUF2	M	24	YBR038w/CHS2	G_2
11	YML091C/RPM2	S	19	YJR092W/BUD4	M	24	YGL116W/CDC20	M
12	YDL127w/PCL2	Late G_1	19	YLR353W/BUD8	M	24	YGR143W/SKN1	M
12	YPR120C/	Late G_1	19	YMR001C/CDC5	M	25	YLR286C/CTS1	Late G_1
12	YDL003W/RHC21	Late G_1	19	YGL021W/ALK1	M	25	YGL089C/MF(alpha)2	Late G_1
12	YAR007C/RFA1	Late G_1	19	YLR131c/ACE2	M	26	YBR200w/BEM1	Early G_1
12	YBL035c/POL12	Late G_1	19	YOR025W/HST3	M	26	YBL023c/MCM2	Early G_1
12	YBR088c/POL30	Late G_1	20	YGR109C/CLB6	Late G_1	26	YBR202w/CDC47	Early G_1
12	YDL164C/CDC9	Late G_1	20	YNL289W/PCL1	Late G_1	26	YEL032w/MCM3	Early G_1
12	YML102W/	Late G_1	20	YLR313C/	Late G_1	26	YLR395C/COX8	Early G_1
12	YPR175W/DPB2	Late G_1	20	YPR018W/RLF2	Late G_1	26	YMR256c/COX7	Early G_1
12	YDR097C/	Late G_1	20	YPL153C/SPK1	Late G_1	26	R281W/YOR1	Early G_1
12	YLR032w/RAD5	Late G_1	20	YBR070c/	Late G_1	26	YOR316C/COT1	Late G_1
12	YML027W/YOX1	Late G_1	21	YJR159W/SOR1	G_2	26	YCR042c/TSM1	M
12	YMR179W/SPT21	Late G_1	21	YBR104w/YMC2	G_2	26	YOR229W/	M
13	YJL187C/SWE1	Late G_1	21	YLR014c/PPR1	G_2	26	YDL138W/RGT2	M
13	YPL256C/CLN2	Late G_1	21	YOR274W/MOD5	G_2	26	YIL167W/	M
13	YMR076C/PDS5	Late G_1	21	YDR464w/SPP41	G_2	27	YNR016C/ACC1	Early G_1
13	YER070w/RNR1	Late G_1	21	YLL046c/RNP1	G_2	27	YBR160w/CDC28	Late G_1
13	YLR103c/CDC45	Late G_1	22	YER111c/SWI4	Early G_1	27	YBR252w/DUT1	Late G_1
13	YNL102W/CDC17	Late G_1	22	YOR373W/NUD1	Early G_1	27	YBR278w/DPB3	Late G_1
13	YOR074C/CDC21	Late G_1	22	YKL092C/BUD2	Early G_1	27	YDR297w/SUR2	Late G_1
13	YKL113C/RAD27	Late G_1	22	YMR199W/CLN1	Late G_1	27	YDL155W/CLB3	S
13	YLR383W/	Late G_1	22	YKL042W/	Late G_1	27	YFR037C/	S
13	YML060W/OGG1	Late G_1	22	YLR212C/TUB4	Late G_1	27	YPL016W/SWI1	S
13	YIL140W/SRO4	S	22	YPL241C/CIN2	Late G_1	27	YDL093W/PMT5	S
13	YAR008W/	S	22	YDR507c/GIN4	Late G_1	27	YKR001C/SPO15	S
14	YDL181W/INH1	Early G_1	22	YGL027C/CWH41	Late G_1	27	YER016w/BIM1	S
14	YML110C/	Early G_1	22	YJL173C/RFA3	Late G_1	27	YER017c/AFG3	S
14	YIL009W/FAA3	Early G_1	22	YNL262W/POL2	Late G_1	27	YPR111W/DBF20	G_2
14	YBR083w/TEC1	Early G_1	22	YDL101C/DUN1	Late G_1	27	YOR188W/MSB1	G_2
14	YPL187W/MF(alpha)1	Late G_1	22	YLR234W/TOP3	Late G_1	27	YJL092W/HPR5	G_2
14	YJR148W/TWT2	Late G_1	22	YML021C/UNG1	Late G_1	27	YKL032C/IXR1	G_2
15	YPL127C/HHO1	Late G_1	22	YLR382C/NAM2	Late G_1	28	YMR190C/SGS1	S
15	YLL021w/SPA2	Late G_1	22	YJL196C/	Late G_1	28	YIR017C/MET28	S
15	YBL063w/KIP1	S	22	YBR073w/RDH54	Late G_1	28	YHR086w/NAM8	S
15	YDR113c/PDS1	S	22	YKL101W/HSL1	Late G_1	28	YJR137C/	S
15	YDR356w/NUF1	S	22	YKL165C/	Late G_1			

4   Conclusion

We propose a method to cluster gene expression temporal profile microarray data. On two well-known real-life datasets, we have demonstrated that using agglomerative clustering with the proposed profile alignment method for measuring similarity produced very good results when compared to that of the Pearson and Spearman correlation similarity measures. We also propose a variant of the $\mathcal{I}$-index that can make a trade-off between minimizing the number of useful clusters and keeping the distinctness of individual clusters. We have also shown the biological significance of the results obtained by computing the classifrcation accuracy of our method in pre-clustered yeast data - the accuracy was over 83%.

PAAC can be used for effective clustering of gene expression temporal profile microarray data. Although we have shown the effectiveness of the method in microarray time-series datasets, we are planning to investigate the effectiveness of the method as well in dose-response microarray datasets, and other microarray time-series data.

Additionally, further studies are desirable to develop an appropriate method that can efficiently perform alignment of multiple profiles so that a cluster mean can be computed after aligning all profiles in a cluster at the same time, and hence removing the effect of ordering of profiles during the alignment. Also, the effects of alignment in the partitional iterative methods (e.g., EM, regression, k-means, SOM, etc.) are yet to be investigated.

Acknowledgements. This research has been partially supported by the Chilean National Council for Technological and Scientific Research, FONDECYT grant No. 1060904. The authors would like to thank Mr. K. Raiyan Kamal, Department of Computer Science and Engineering, Bangladesh University of Engineering and Technology (BUET), for his help with some of the computations performed for this paper.

References

1. Bari, A., Rueda, L.: A New Profile Alignment Method for Clustering Gene Expression Data. In: Lamontagne, L., Marchand, M. (eds.) Canadian AI 2006. LNCS, vol. 4013, pp. 86–97. Springer, Heidelberg (2006)
2. Bréhélin, L.: Clustering Gene Expression Series with Prior Knowledge. In: Casadio, R., Myers, G. (eds.) WABI 2005. LNCS (LNBI), vol. 3692, pp. 27–38. Springer, Heidelberg (2005)
3. Cho, R.J., Campbell, M.J., Winzeler, E.A., Steinmetz, L., Conway, A., Wodicka, L., Wolfsberg, T.G., Gabrielian, A.E., Landsman, D., Lockhart, D.J., Davis, R.W.: A genome-wide transcriptional analysis of the mitotic cell cycle. Molecular Cell 2(1), 65–73 (1998)
4. Conesa, A., Nueda, M.J., Ferrer, A., Talon, M.: maSigPro: a method to identify significantly differential expression profiles in time-course microarray experiments. Bioinformatics 22(9), 1096–1102 (2006)
5. Chu, S., DeRisi, J., Eisen, M., Mulholland, J., Botstein, D., Brown, P., Herskowitz, I.: The transcriptional program of sporulation in budding yeast. Science 282, 699–705 (1998)

122   L. Rueda, A. Bari, and A. Ngom

6. Déjean, S., Martin, P.G.P., Baccini, A., Besse, P.: Clustering Time-Series Gene Expression Data Using Smoothing Spline Derivatives. EURASIP J. Bioinform. Syst. Biol. 2007, 70561 (2007)
7. Drăghici, S.: Data Analysis Tools for DNA Microarrays. Chapman & Hall, Boca Raton (2003)
8. Eisen, M., Spellman, P., Brown, P., Botstein, D.: Cluster analysis and display of genome-wide expression patterns. Proc. Natl. Acad. Sci. 95, 14863–14868 (1998)
9. Ernst, J., Nau, G.J., Bar-Joseph, Z.: Clustering Short Time Series Gene Expression Data. Bioinformatics 21(suppl. 1), i159–i168 (2005)
10. Gasch, A.P., Eisen, M.B.: Exploring the conditional coregulation of yeast gene expression through fuzzy k-means clustering. Genome Biology 3(11), 0059.1–0059.22 (2002)
11. Guillemin, K., Salama, N., Tompkins, L., Falkow, S.: Cag pathogenicity island-specific responses of gastric epithelial cells to Helicobacter pylori infection. Proc. Natl. Acad. Sci. 99, 15136–15141 (2002)
12. Hartigan, J.A.: Clustering Algorithms. John Wiley and Sons, Chichester (1975)
13. Heijne, W.H., Stierum, R.H., Slijper, M., van Bladeren, P.J., van Ommen, B.: Toxicogenomics of bromobenzene hepatotoxicity: a combined transcriptomics and proteomics approach. Biochem. Pharmacol. 65, 857–875 (2003)
14. Heyer, L., Kruglyak, S., Yooseph, S.: Exploring expression data: identification and analysis of coexpressed genes. Genome Res. 9, 1106–1115 (1999)
15. Hogg, R., Craig, A.: Introduction to Mathematical Statistics, 5th edn. MacMillan, Basingstoke (1995)
16. Hwang, J., Peddada, S.: Confidence interval estimation subject to order restrictions. Ann. Statist. 22, 67–93 (1994)
17. Iyer, V., Eisen, M., Ross, D., Schuler, G., Moore, T., Lee, J., Trent, J., Staudt, L., Hudson Jr., J., Boguski, M.: The transcriptional program in the response of human fibroblasts to serum. Science 283, 83–87 (1999)
18. Bar-Joseph, Z., Gerber, G., Jaakkola, T., Gifford, D., Simon, I.: Continuous representations of time series gene expression data. Journal of Computational Biology 10(3-4), 341–356 (2003)
19. Lobenhofer, E., Bennett, L., Cable, P., Li, L., Bushel, P., Afshari, C.: Regulation of DNA replication fork genes by 17betaestradiol. Molec. Endocrin. 16, 1215–1229 (2002)
20. Maulik, U., Bandyopadhyay, S.: Performance Evaluation of Some Clustering Algorithms and Validity Indices. IEEE Transactions on Pattern Analysis and Machine Intelligence 24(12), 1650–1654 (2002)
21. Moller-Levet, C., Klawonn, F., Cho, K.-H., Wolkenhauer, O.: Clustering of unevenly sampled gene expression time-series data. Fuzzy sets and Systems 152(1,16), 49–66 (2005)
22. Peddada, S., Prescott, K., Conaway, M.: Tests for order restrictions in binary data. Biometrics 57, 1219–1227 (2001)
23. Peddada, S., Lobenhofer, E., Li, L., Afshari, C., Weinberg, C., Umbach, D.: Gene selection and clustering for time-course and dose-response microarray experiments using order-restricted inference. Bioinformatics 19(7), 834–841 (2003)
24. Petrie, T.: Probabilistic functions of finite state Markov chains. Ann. Math. Statist. 40, 97–115 (1969)
25. Ramoni, M., Sebastiani, P., Kohane, I.: Cluster analysis of gene expression dynamics. Proc. Natl. Acad. Sci. USA 99(14), 9121–9126 (2002)
26. Ramsay, J., Silverman, B.: Functional Data Analysis, 2nd edn. Springer, New York (2005)

27. Rueda, L., Bari, A.: Clustering Temporal Gene Expression Data with Unequal Time Intervals. In: 2nd International Conference on Bio-Inspired Models of Network, Information, and Computing Systems, Bioinformatics Track, Budapest, Hungary (2007) ICST 978-963-9799-11-0

28. Schliep, A., Schonhuth, A., Steinhoff, C.: Using hidden Markov models to analyze gene expression time course data. Bioinformatics 19, I264–I272 (2003)

29. Spellman, P.T., Sherlock, G., Zhang, M.Q., Iyer, V.R., Anders, K., Eisen, M.B., Brown, P.O., Botstein, D., Futcher, B.: Comprehensive identification of cell cycleregulated genes of the yeast saccharomyces cerevisiae by microarray hybridization. Mol. Biol. Cell. 9, 3273–3297 (1998)

30. Tamayo, P., Slonim, D., Mesirov, J., Zhu, Q., Kitareewan, S., Dmitrovsky, E., Lander, E., Golub, T.: Interpreting patterns of gene expression with self-organizing maps: Methods and application to hematopoietic differentiation. Proc. Natl. Acad. Sci. 96(6), 2907–2912 (1999)

31. Tavazoie, S., Hughes, J., Campbell, M., Cho, R., Church, G.: Systematic determination of genetic network architecture. Nat. Genet. 22, 281–285 (1999)

32. Zhu1, G., Spellman, P.T., Volpe, T., Brown, P.O., Botstein, D., Davis, T.N., Futcher, B.: Two yeast forkhead genes regulate cell cycle and pseudohyphal growth. Nature 406, 90–94 (2000)

Integrating Thermodynamic and Observed-Frequency Data for Non-coding RNA Gene Search

Scott F. Smith[1] and Kay C. Wiese[2]

[1] Electrical and Computer Engineering Department, Boise State University,
Boise, Idaho, 83725, USA
SFSmith@BoiseState.edu
[2] School of Computing Science, Simon Fraser University,
Surrey, BC, V3T 0A3, Canada
Wiese@cs.sfu.ca

Abstract. Among the most powerful and commonly used methods for finding new members of non-coding RNA gene families in genomic data are covariance models. The parameters of these models are estimated from the observed position-specific frequencies of insertions, deletions, and mutations in a multiple alignment of known non-coding RNA family members. Since the vast majority of positions in the multiple alignment have no observed changes, yet there is no reason to rule them out, some form of prior is applied to the estimate. Currently, observed-frequency priors are generated from non-family members based on model node type and child node type allowing for some differentiation between priors for loops versus helices and between internal segments of structures and edges of structures. In this work it is shown that parameter estimates might be improved when thermodynamic data is combined with the consensus structure/sequence and observed-frequency priors to create more realistic position-specific priors.

Keywords: Bioinformatics, Covariance models, Non-coding RNA gene search, RNA secondary structure, Database search.

1 Introduction

The present decade has seen a surge of interest in and knowledge about non-coding RNA (ncRNA) [1]. These RNA molecules perform some function which generally depends on the molecule's shape instead of or in addition to acting as a carrier of information. Since molecular structure rather than sequence tends to be much more highly conserved, homology search based solely on sequence typically does not give good results. However, knowledge of intra-molecular base pairing (secondary structure) generally is sufficient and full three-dimensional modeling of molecular structure is typically not necessary for good homology search results.

An additional difference between searching for ncRNA genes versus the more highly developed protein-coding gene search methods is that it is generally not possible to first search for ncRNA genes generically and then classify these putative genes into families in a second step [2]. Protein-coding genes usually have strong signals

C. Priami et al. (Eds.): Trans. on Comput. Syst. Biol. X, LNBI 5410, pp. 124–142, 2008.
© Springer-Verlag Berlin Heidelberg 2008

such as the presence of open reading frames or variation in nucleotide composition which allow for generic protein-coding gene search [3]. Gene search programs for ncRNA are generally applied to all the genomic data repeatedly for each ncRNA family. In addition to this causing computational overhead issues, the larger amount of data processed increases the number of false positives generated. Keeping the number of false positives manageable by keeping detection thresholds high increases the need for search algorithms with high discriminatory power between true genes and background noise.

Covariance models are a very powerful ncRNA gene search tool. They are, unfortunately, also extremely slow (an issue addressed elsewhere [4][5]). A covariance model (CM) can be thought of as an extension of a hidden-Markov model (HMM) [6]. HMMs allow for position-specific insertion, deletion, and mutation probabilities relative to the consensus sequence of a gene family. They are a powerful, but slow, method for homology search based on sequence alone. CMs allow for consensus secondary structure information to be added to the HMM.

A very commonly used implementation of CMs for ncRNA gene search is the Infernal [7] package. This package contains programs to estimate covariance model parameters from a set of known ncRNA gene family members and to use these parameters to search for new family-member genes in genomic data (among other things). The package is the basis for the popular Rfam RNA families and models database [8] (much as the HMMer [9] package forms the basis for Rfam's sister, the Pfam protein families and models database [10]). Infernal has recently incorporated the ability to have informative priors [11] for insertion, deletion, and mutation probabilities based on observed frequencies of events in the overall Rfam database (including non-family-member data). The priors differentiate between positions in loops and helices and between positions within loops or helices and positions at the ends of these structures. They do not, however, take into account information such as the consensus loop length, whether the loop is of the hairpin, internal, or bulge type, or the consensus composition of neighboring helix base pairs to determine the likelihood of helix base-pair mutations. The research presented here is designed to address these shortcomings and to include more independent information into the parameter estimates.

This paper builds on the ideas presented in [20]. Whereas the former work was conceptual in nature, the present work includes detailed examination of a real ncRNA family (Section 2.2) and expands on the concepts previously presented.

1.1 Covariance Models

Covariance models are composed of an interconnection of nodes forming a tree [12]. Table 1 shows the types of nodes that may appear in a covariance model tree. The tree is rooted at a node called the *root start node* (root S node) which is assigned node index 0. All branches of the tree are terminated by nodes of the *end node* (E node) type. End nodes represent a null subsequence of nucleotides on which larger subsequences may be built. Each node in the covariance model is the root of a subtree representing a contiguous subsequence of the ncRNA family consensus sequence. The three emitting CM node types (L, R, and P) take the consensus subsequence of their child node and add an additional consensus position to the left (L node), right (R node), or both left and right (P node). All nodes other than E and B type nodes have exactly one child. E type nodes have no children and *bifurcation nodes* (B nodes) have exactly two children. The B nodes serve to join two subsequences into a single contiguous subsequence.

In order to allow for insertions and deletions relative to the consensus sequence, covariance model nodes have an internal state structure. All nodes have a single internal state that matches the purpose of the overall CM node (called the consensus state in Table 1). For instance, an ML state used as a consensus state emits a single consensus nucleotide to the left of a child subsequence. Many CM node types also have additional non-consensus states that allow for insertions or deletions. Two of these non-consensus states (ML and MR) behave exactly as their consensus counterparts and are therefore not given separate nomenclature. Table 2 shows the purposes of the five possible non-consensus states. The MP, B, E, and S state types are only used as consensus states and never as non-consensus states. It is the MP state type that is the main differentiator between the CM and the HMM, since this state allows simultaneous addition of both nucleotides of a base pair with a joint probability of emission specified rather than two marginal probabilities if ML and MR were used separately (both ML and MR would be just M states in an HMM).

Table 1. Node types of covariance models

Node type	Description of node purpose	Consensus state	Non-consensus states
L	Add a nucleotide to left end of subsequence	ML	D, IL
R	Add a nucleotide to right end of subsequence	MR	D, IR
P	Simultaneously add one nucleotide to each end	MP	D, IL, IR, ML, MR
B	Join two contiguous subsequences (bifurcation)	B	none
E	Create null subsequence to build on	E	none
S (root)	Finalize scoring of complete sequence	S	IL, IR
S (left)	Finalize scoring of a left bifurcation child	S	none
S (right)	Finalize scoring of a right bifurcation child	S	IL

Table 2. Internal non-consensus state types for covariance model nodes

State type	Description of state purpose
D	Do not include consensus position (L or R node) or positions (P node)
IL	Add one or more non-consensus nucleotides to left of subsequence (before adding consensus nucleotides)
IR	Add one or more non-consensus nucleotides to right of subsequence (before adding consensus nucleotides)
ML	Include only the left member of the consensus pair in a P node
MR	Include only the right member of the consensus pair in a P node

Each state in the CM that may emit nucleotides (states of type MP, ML, MR, IL, and IR) has a set of emission probabilities (4 probabilities if one nucleotide is emitted and 16 probabilities if two nucleotides are emitted). All states also have transition probabilities to all states with which they are connected. States may only be connected to other states in the same node or in a parent or child node. Not all possible connections of states within the node group comprised of a node, its parent, and its children are made.

Scoring of a sequence against the covariance model is done with dynamic programming. A score of 0 is assigned to the E states in the E nodes. The scores at each of the remaining states in the model can be determined from the scores of each of the child states, the transition probabilities from each of these children, and the emission probabilities. The child state which maximizes the score at a state (through the child state's score and child's transition probability) is chosen to calculate the parent state's score. The score of interest is that of the root start state (in the root start node), but all other state scores must be calculated to get the root start state score.

1.2 Parameter Estimation Using Family Member Observed Frequencies Only

A covariance model of an ncRNA family can be generated from a structure-annotated multiple alignment of known family member sequences. Each column of the multiple alignment must be chosen as either conserved or non-conserved. The conserved columns are assigned to emitting covariance model node types (L, R, or P). The annotation of the consensus structure consists of labelling the conserved columns of the multiple alignment as one of three types: unpaired, paired with a column further to the right, or paired with a column further to the left. If the secondary structure does not contain pseudoknots, then this structure labelling is not ambiguous. If pseudoknots do exist, it is necessary to treat some of the base-paired positions as if they were unpaired since covariance models can not handle pseudoknots. The unpaired columns should all correspond to L or R nodes and the paired columns with P nodes.

Gaps in the multiple alignment in conserved columns are deletions and the probability of a deletion at a specific position in the consensus sequence can be estimated as the ratio of the number of observed gaps to the number of sequences in the alignment. For L and R nodes, this probability should be reflected in the relative transition probabilities in and out of the D state compared to the transition probabilities in and out of the consensus (ML or MR) state. For P nodes, both columns of the base pair must be considered. The observed frequency of gaps in both columns leads to the transition probabilities for the D state relative to the MP state. The observed frequency of gaps in the left column only determines the relative probabilities for the MR and MP states and gaps in the right column only determines relative probabilities for ML versus MP.

Non-gaps in the multiple alignment non-conserved columns are insertions relative to the consensus sequence. The observed frequency of insertions (of any length) between two conserved columns determines the probability of visiting the IL or IR state relative to simply transitioning from child-node consensus state to parent-node consensus state. Insert type states (IL and IR) are the only states with a connection to themselves. The observed frequency of insertions at a position of length two or more relative to the observed total number of insertions at that position determines the self-transition probability.

For single emission states (ML, MR, IL, and IR) the relative frequencies of the four nucleotides (A, C, G, and U) in the multiple alignment column allow calculation of the emission probabilities. Counts of each of the sixteen possible nucleotide pairs result in MP state emission probabilities.

A significant problem with this position-specific observed-frequency approach is that usually almost all of the observed frequencies are 0. If there is no theoretical reason to rule out a particular insertion, deletion, or mutation, then the count of 0 is likely a result of the sample of known sequences being too small. It is possible to get around this problem through the use of priors. The simplest prior is to add a single pseudocount to each possibility such that all estimated probabilities are more than 0, but remain small if the event was not actually observed. A method which is currently in use for determining better priors using data from non-family sequences is discussed in the next subsection. The remainder of this paper discusses further improvement using thermodynamic data and the consensus sequence/structure.

1.3   Parameter Estimation Including Non-family Observed Frequencies

More informative priors for state transitions and emissions than those offered by using uniform pseudocounts might be obtained using generic (non-family) observed-frequency data from a ncRNA sequence database. The uniform pseudocount method basically includes the non-family information that nothing is impossible (which is arguably more true than the alternative that anything not observed in known sequences is impossible). Better non-family information would include insertion, deletion, and mutation probabilities that vary depending on the type of secondary structure in which the sequence position resides. One might suppose, for instance, that an insertion between two non-base-paired positions would be more likely than between two base-paired positions, since the former represents an insertion in some type of loop structure, whereas the later disrupts a helix structure. Priors such as these have recently been implemented in the Infernal ncRNA analysis package (starting with version 0.6, November 2005) and applied in the creation of the Rfam version 8.0 database (February 2007). These priors are calculated using a Dirichlet mixture model [13].

Priors for transition probabilities depend on the types of states transitioned into and out of (MP, IL, etc.), the type of node which contains the state transitioned into, and the type of node which is the child of the node containing the state transitioned into. There is always exactly one child node when the node type is anything other than B or E. E-type nodes contain only E states which have no transition in, so there is no transition prior. B-type nodes have two children, but the transition probabilities are both 1 by definition, so there is no reason to calculate a prior. Emission priors depend only on whether the state is single-emission (ML, MR, IL, IR) or pair-emission (MP).

The dependence of transition priors on node type and child node type clearly allows differentiation between insertions and deletions in helices versus loops. However, there is little discrimination between different types of loops (bulge, internal, hairpin) and no discrimination between different size loops (for example, tight hairpin loops versus large hairpin loops). Mutation probabilities in helices also do not depend on the consensus nucleotides in neighboring base pairs, even though thermodynamic data show that helix stacking energies do depend on neighboring base pairs [14]. In general, little use is made of the consensus sequence or structure information to guide the choice of transition or emission priors.

2 Information in Consensus Loop Lengths and Types

Figures 1 and 2 show two example structures taken from the Rfam database (version 8.0). Figure 1 shows a bulge loop on the left side (5' side) of the only helix in the Cardiovirus CRE [15] ncRNA gene family (Rfam accession number RF00453). This family happens to have no known members with insertions or deletions relative to the consensus sequence. It is unclear whether the lack of known insertions or deletions is a result of there not being any to find or if actual members have not yet been found because the insertion and deletion penalties in the covariance model are too severe to allow scoring above the search threshold. Since there are no known insertions or deletions, the model transition probabilities reflect only the Dirichlet mixture priors used in Infernal. The transition scores used in the model depend on the number of known sequences used to build the model (since the observed frequencies are given a larger weight relative to the priors when more known sequences are used to build the model), but all transitions between the same state types with the same parent-state node type and parent-state child node type must have the same transition score within the model. The Cardiovirus CRE model was built using 12 seed sequences. The exact weighting between prior and observation is further complicated by the use of entropy weighting to determine the effective number of observed sequences. If many very similar known sequences are used to build the model, then these sequences are treated as if they were smaller in number than if the sequences had more variability.

The 3-nucleotide bulge loop in Cardiovirus CRE is at consensus sequence positions 8 through 10 and is modeled with three L nodes in the covariance model which just happen to also have node indices 8 through 10. Position-specific penalties for single insertions or deletions relative to the consensus sequence can be determined from the covariance model parameters files and are shown in Figure 1 for the four

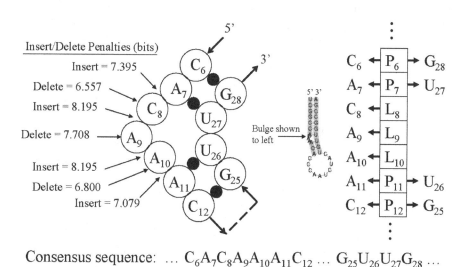

Consensus sequence: ... $C_6A_7C_8A_9A_{10}A_{11}C_{12}$ ... $G_{25}U_{26}U_{27}G_{28}$ ...

Fig. 1. Bulge loop in Cardiovirus CRE (RF00453)

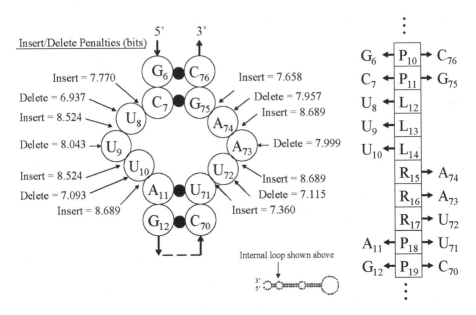

Consensus sequence: ... $G_6C_7U_8U_9U_{10}A_{11}G_{12}$ ... $C_{70}U_{71}U_{72}A_{73}A_{74}G_{75}C_{76}$ ...

Fig. 2. Internal loop in mir-16 (RF00254)

possible insertion positions and three possible deletion positions associated with the bulge loop. These penalties are calculated as the loss of score (in bits) resulting from changes in state transitions necessary to map the consensus sequence with a single insertion or deletion to the model relative to mapping the consensus sequence itself to the model. Two of the insertion penalties (both 8.195) have to be the same since they involve consensus-to-consensus, consensus-to-insert, and insert-to-consensus transitions where both the IL-state-containing node and IL-state child node are of the L type. The delete penalties can be different since states in three nodes are involved in the penalty calculation. Even though most of the penalties are capable of being different in such a short loop structure, the actual variation shown in Figure 1 is not very large.

A second example from Rfam shown in Figure 2 is that of the mir-16 [16] ncRNA gene family (RF00254). This family contains an internal loop with three nucleotides on both the left side (5' side) and right side (3' side). There are no known insertions or deletions in mir-16, just as in Cardiovirus CRE. Two of the insert penalties on each side of the internal loop must be the same (8.524 on the left side, and 8.689 on the right side) for the same reason that two of the Cardiovirus CRE insert penalties had to be the same. The variation of the penalties is small from position to position within the mir-16 internal loop. What is striking, is that the penalties are not very different between the internal loop and the bulge loop. In particular, insertions in the bulge loop would seem much more disruptive than insertions in the internal loop as will be indicated by the thermodynamic data presented below. However, it is not surprising that the values are so similar, since the priors are estimated from a mixture of bulge, internal, hairpin, and other loops of various lengths in the general database. Consensus secondary structure information is not used to determine choice of prior.

2.1 Potential Use of Thermodynamic Data for Improvement of Loop Priors

There is a long history of use of experimentally determined free energy changes from the formation of RNA structures in the prediction of RNA secondary structure from sequence [17-18]. It is not unreasonable to believe that knowledge of which changes to sequence result in radical changes in molecular stability and which do not might help determine which changes might be evolutionarily conserved and which not. This information might be especially helpful in cases where there are few or no direct observations of sequence changes. This subsection investigates use of measured free energy changes for formation of loops of various lengths and types for improvement of state transition priors for covariance model states in and surrounding loop structures.

The three loop structures that are investigated are hairpin loops, bulge loops, and internal loops. Hairpin loops are a contiguous sequence of unpaired nucleotides which connect at each end to the same helix terminus. Bulge loops are single sequences of unpaired nucleotides interrupting a helix structure on only one side of the helix. Internal loops are composed of two non-contiguous sequences that interrupt a helix structure on both sides of the helix between the same two base pairs. Internal loops can be though of as simply a pair of bulges and, in fact, this is exactly how a covariance model handles them. As will be seen, treating an internal loop as two independent bulge loops may result in the loss of important information. Multi-loops (loops with more than two attached helix structures), unstructured connectors, and dangling ends, which all also involve unpaired positions are not considered here.

Table 3 shows measured free energies of formation for the three types of loops (hairpin, bulge, and internal) and for various lengths of the loop sequences [14]. Internal loops are composed of two sequences and length means the sum of the two sequence lengths. An internal loop of length 1 is not possible, since in order to have two sequences both must be at least of length 1. (An internal loop of length 1 on one side and 0 on the other is actually a bulge loop.) Hairpin loops of length 1 or 2 can not be formed since the turn is too tight and hence the free energy of formation can not be measured (and can be thought of as being equal to plus infinity). The table also shows the changes in free energy when the loop length is increased by 1 as the result of a single insertion in the loop. Free energy changes for deletions are just the negative of those for insertion, where the starting length is 1 greater. In general, inserting a nucleotide into a bulge loop is more thermodynamically disruptive than inserting into an internal loop. There is also a preferred hairpin loop length of about 7 to 8 where either insertions or deletions in the hairpin loop increase free energy. Clearly, the effect on molecular stability of insertions or deletions in loop structures depends on loop type and length.

The consensus loop type and length are known at the time state transition priors are used to estimate the covariance model parameters. It is not immediately clear, however, how the free energy changes (in units of kcal/mol) can be used to calculate state transition priors (in units of bits). One possibility is to regress free energy changes on

Table 3. Free energy increments for formation of internal, bulge, and hairpin loops of various lengths (kcal/mol) [14] and free energy changes resulting from single insertions in those structures

Loop length	Formation of internal loops	Change for internal insertion	Formation of bulge loops	Change for bulge insertion	Formation of hairpin loops	Change for hairpin insertion
1	not possible	(bulge loop)	+3.3	+1.9	too tight	N/A
2	+0.8	+0.5	+5.2	+0.8	too tight	N/A
3	+1.3	+0.4	+6.0	+0.7	+7.4	-1.5
4	+1.7	+0.4	+6.7	+0.7	+5.9	-1.5
5	+2.1	+0.4	+7.4	+0.8	+4.4	-0.2
6	+2.5	+0.1	+8.2	+0.9	+4.3	-0.2
7	+2.6	+0.2	+9.1	+0.9	+4.1	0.0
8	+2.8	+0.3	+10.0	+0.5	+4.1	+0.1
9	+3.1	+0.5	+10.5	+0.5	+4.2	+0.1
10	+3.6	+0.4	+11.0	+0.4	+4.3	+0.3
12	+4.4	+0.35	+11.8	+0.35	+4.9	+0.35
14	+5.1	+0.25	+12.5	+0.35	+5.6	+0.25
16	+5.6	+0.3	+13.0	+0.3	+6.1	+0.3
18	+6.2	+0.4	+13.6	+0.2	+6.7	+0.2
20	+6.6	+0.2	+14.0	+0.2	+7.1	+0.2
25	+7.6	+0.16	+15.0	+0.16	+8.1	+0.16
30	+8.4	-	+15.8	-	+8.9	-

observed frequencies of insertion and deletion events, then use the regression coefficients to calculate priors where no observed insertions or deletions exist. This still leaves open the question of functional form in the regression equation. One could simply use free energy change as the regression variable with the expectation that lower free energy is always good. A second possibility is that the absolute value of the free energy change should be the regression variable since larger absolute free energy changes are likely associated with larger changes in overall three-dimensional shape of the molecule (including the possibility that the molecule is simply not stable). Other possibilities are that some non-linear function such as square, square root, or logarithm should be taken. The observed-frequency priors are themselves the result of taking a logarithm (and hence the units of bits).

Rather than try to use free energy changes to try to predict insertion and deletion frequencies and then combine predicted frequencies with observed frequencies, it is also possible to predict insertion or deletion penalties using free energy change directly. A regression of penalties (in bit units) at positions with known insertions or deletions on free energy change (or absolute free energy change) results in a predicted penalty (a prior in units of bits) for positions without known insertions or deletions if the consensus free energy change is known (which it is).

There is already a slight dependence of observed-frequency priors as used by Infernal on loop type (but not length) due to the use of enclosing node and child node type in prior selection. Table 4 shows this dependence. The loop ends are treated differently than loop interiors as a result of the relevant insertion state being enclosed in a P-type node or the child node being a P-type or E-type node. Hairpin loops

Table 4. Observed-frequency insertion prior usage by loop type and position within loop

Insert position	CM node type	CM state type	CM child node type	Loop types [a]
Before leftmost	P	IL	L	Hairpin, Left bulge, Internal (left side)
Before leftmost	R	IR	P	Right bulge, Internal (right side)
Interior	L	IL	L	Hairpin, Left bulge, Internal (left side)
Interior	R	IR	R	Right bulge, Internal (right side)
After rightmost	L	IL	E	Hairpin [b]
After rightmost	P	IR	L	Hairpin [b], Internal (right side)
After rightmost	L	IL	P	Left bulge
After rightmost	P	IR	R	Right bulge
After rightmost	L	IL	R	Internal (left side)

a) Hairpin loops are modeled as a string of L nodes bracketed by an E node below and a P node above. Left bulges (on the 5' side of a helix) are strings of L nodes with P nodes above and below. Right bulges (on the 3' side of the helix) are strings of R nodes with P nodes above and below. Internal loops are modeled as a string of L nodes (the 5' side of the loop) above a string of R nodes (the 3' side of the loop) all bracketed by P nodes above and below. b) Insertions after the rightmost (closest of 3' end) position in a hairpin loop can be done two ways, either by the L node representing the rightmost consensus position of the hairpin loop or by the enclosing P node. Choosing the insertion penalty very large for the IR state of the P node would force the IL state of the L node to determine the effective prior.

are always modeled as a series of L-type nodes by convention and the E state always specifies the null sequence between the right end of the hairpin loop and the start of the right half of the enclosing helix structure. Bulge loops on the left half of a helix require modeling by L nodes and bulge loops on the right half require R-node modeling. Internal loops are composed of a right bulge and a left bulge. By convention, the left bulge L nodes are placed closer to the covariance model tree root and the right bulge R nodes further from the root (see Figure 2). There is a small ambiguity in the covariance model node definitions in that two different insertion states are capable of inserting where the E state of the hairpin loop is located. An L-node IL state can insert to the left of the E state and a P-node IR can insert to the right of the E state, but since the E state specifies a null sequence the two insertions are indistinguishable. This can be resolved by always making the P-node insertion penalty very large compared to that of the L-node such that the L-node transition probabilities determine the overall insertion probability. Unfortunately, examination of Rfam models shows that the P-node insertion penalty is usually lower and therefore determines the effective probability.

For the rightmost position there could be perfect separation between hairpin, bulge, and internal loop types. For all other positions, these three loop types are mixed together and any prior estimated from the overall database will be a mixture of observations from the three different loop types. The thermodynamic data in Table 3 indicates that this might not be the best situation. Furthermore, there is no dependence of the priors on consensus loop lengths.

2.2 Evidence That Thermodynamic Data Helps Predict Insertion Probability

The Rfam database contains both multiple alignments and covariance model parameter files for ncRNA families that now number over 600. In order to see if there is any evidence that inclusion of thermodynamic data into covariance model prior estimates might be helpful, one of these ncRNA families is examined in detail with respect to insertion events. The U4 family (Rfam accession number RF00015) [19] was chosen since it is well studied, contains a number of loop structures including at least one of each type (hairpin, bulge, and internal) and has a fairly large number of gaps in its seed multiple alignment.

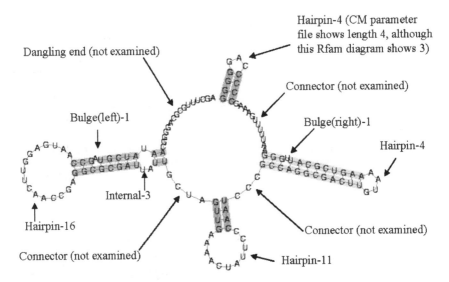

Fig. 3. Consensus secondary structure of U4 (RF00015) ncRNA family

All unpaired consensus model positions in the U4 family are examined that correspond to either a hairpin, bulge, or internal loop. Figure 3 shows the consensus structure of this family. The consensus structure contains four hairpin loops (of lengths 4, 4, 11, and 16), two bulge loops each of length 1 (one on the left side of a helix and one on the right side of a helix), and an internal loop of length 3.

Insertion events are defined here as one or more insertions with respect to the consensus sequence at one or more locations within a range of sequence positions. Insertions occur between consensus sequence positions, but it is sensible to talk about an insertion occurring at a position if one means that the CM node which models the consensus position also contains the insertion state responsible for a particular insertion location. This definition is used here. Since all insertions are done within a node before any consensus nucleotides are added, P-type nodes are often responsible for unpaired insertions on the left or right end of a loop sequence. Table 5 shows observed insertion events in the 30 seed sequences used to estimate the parameters of the U4 covariance model. Event counts have been made separately for loop ends versus internal segments of loops since Infernal allows different priors for these insertions.

Table 5. Observed loop insertions in 30 seed sequences of U4 ncRNA (RF00015)

Structure at insertion location	Consensus length of structure	Consensus sequence position [b]	Observed insertion events [c]	Observed insertion event rate	CM node-state types for insertion	CM child node type
Internal	3	18 [a]	0	0	P - IL	L
		19	0	0	L - IL	P
		54 [a]	0	0	P - IR	R
		53	5	0.167	R - IR	R
		52	0	0	R - IR	P
Bulge	1	24 [a]	0	0	P - IL	L
		25	0	0	L - IL	P
Hairpin	16	28 [a]	0	0	P - IL	L
		29-43	11	0.024	L - IL	L
		44	0	0	L - IL	E
Hairpin	11	64 [a]	0	0	P - IL	L
		65-74	10	0.033	L - IL	L
		75	0	0	L - IL	E
Hairpin	4	95 [a]	2	0.067	P - IL	L
		99-99	0	0	L - IL	L
		100	10	0.333	L - IL	E
Bulge	1	111 [a]	0	0	P - IR	R
		110	9	0.300	R - IR	P
Hairpin	4	129 [a]	0	0	P - IL	L
		130-132	7	0.078	L - IL	L
		133	0	0	L - IL	E

a) Helix sequence positions with possibility of insertion between the helix and an adjacent non-helix structure. b) Consensus sequence positions for which the covariance model node assigned to the position contains an insert state capable of generating insertions into the non-helix structure. c) One or more insertions at a possible non-helix insertion location. Insertion of multiple nucleotides at a single position in a single sequence is counted as a single event.

The question here is how well the priors used by Infernal on the state transition probabilities helped predict insertions that are actually observed. To answer the question, one needs to figure out what the transition parameters would have been had no insertions been observed. This can be done by looking at positions in U4 without observed insertions with the same node type and child node type. Table 6 shows the single-insertion penalty for each of the possible combinations of node type, child node type, and inserting state in U4. Insertion penalty is calculated as the difference between the consensus-to-consensus state transition score and the consensus-to-insert plus insert-to-consensus state transition scores. Insertions of length more than one would also add one or more insert-to-insert state transition scores, but these have been ignored to simplify the analysis. As was seen before in Figure 1 (Cardiovirus CRE bulge loop) and Figure 2 (mir-16 internal loop), there is very little variability in the insertion penalty and it is unlikely that these values can predict much of anything. The underlying reason for this is apparent from Table 4. These values are derived from observed insertion events in the overall database which are a mixture of events in different loop types and loop lengths.

Table 6. Observed-frequency single-insertion-penalty priors used by Infernal 0.7 for U4 ncRNA (RF00015) and observed insertion event rates for 30 seed sequences

CM node-state types for insertion	CM child node type	Example node number	Prior single-insert penalty (bits) [a]	Observed insertion rate in U4
P - IL	L	34	8.038	0.022 (4/180)
P - IR	L	34	8.570	0.000 (0/30)
P - IR	R	81	8.063	0.000 (0/30)
L - IL	P	56	7.536	0.000 (0/30)
L - IL	L	49	8.766	0.233 (28/120) [b]
L - IL	E	50	8.860	0.083 (10/120)
L - IL	R	49	8.732	0.000 (0/30)
R - IR	P	25	7.572	0.150 (9/60)
R - IR	R	none	-	0.167 (5/30)

a) Taken from positions in RF00015 covariance model parameter file at positions where no seed alignment insertions are observed. Calculated as $t(cc\text{-}pi) + t(pi\text{-}pc) - t(cc\text{-}pc)$, where $t(cc\text{-}pi)$ is the transition score from the child node consensus state to the parent node insert state, $t(pi\text{-}pc)$ is the transition score from the parent insert state to the parent consensus state, and $t(cc\text{-}pc)$ is the transition score from the child consensus state to the parent consensus state. b) IL (insert left) states in L (left consensus emission) nodes with L child nodes appear only in the interior (non-helix-adjacent) positions of the four hairpin loops of U4. This is the only node/state/child combination for which multiple adjacent consensus sequence positions have the same combination. An insertion of one or more nucleotides at one or more positions in the interior of the hairpin loop is taken as a single event and the loop interior is counted as only one opportunity for an insertion event.

Table 7. Change in free energy for single insertions based on consensus structure of U4 ncRNA (RF00015) and observed rates of insertion events

Structure	Consensus length	Insertion ΔG (kcal/mol)	Observed insertion rate
Internal	3	+0.4	0.167 (5/30) [a]
Bulge	1	+1.9	0.150 (9/60) [b]
Hairpin	16	+0.3	0.367 (11/30)
Hairpin	11	+0.3	0.333 (10/30)
Hairpin	4	-1.5	0.150 (9/60) [b]

a) Values in parentheses are the number of event observations and the number of seed sequences times the number of occurrences of the structure. An insertion event is an insertion of length one or more at one or more positions within the structure. A given structure in a given sequence may have zero or one insertion events, but not more. b) There are two Bulge-1 and two Hairpin-4 structures in U4.

As an alternative, one could try to predict insertion rates based on some function of the change in free energy expected from a single insertion based on thermodynamic measurement data (Table 3). In Table 7, the free energy changes for single insertions in each of the consensus loop types and lengths found in U4 are shown. These do not depend on whether the insertion is at a loop end or interior to the loop. One sees that there seems to be a tendency for insertions with a low absolute change in free energy to have higher observed insertion rates.

In Figure 4, the observed insertion rates from Table 6 are plotted against the prior single-insert penalty. There is no tendency for the prior penalty to be larger for sequence locations that subsequently had higher rates of insertion activity. A regression line on this plot would actually have positive slope (when it should be negative), but would be statistically the same as zero slope at any reasonable level of statistical significance.

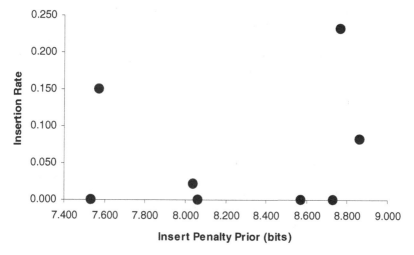

Fig. 4. Loop Insert Penalty Priors versus Observed Insertion Rates in U4

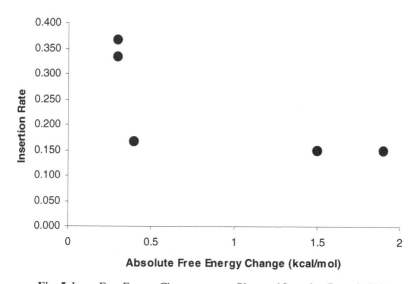

Fig. 5. Loop Free Energy Changes versus Observed Insertion Rates in U4

Figure 5 plots insertion rate versus the absolute value of the free energy change from Table 7. There appears to be a decrease in insertion rate as the free energy deviates more from its consensus value. However, this relation does not seem to be linear. There are not enough data points to do any meaningful further analysis such as attempting to determine the probable functional relationship. While Figure 5 is indicative of possible predictive power for insertion probabilities by absolute free energy changes, it is not very useful in this form. What is needed is a prior in units of bits that can be combined with observed insertions in known family members to estimate covariance model parameters.

Figure 6 regresses the insertion penalties at loop positions in U4 that have observed insertions on absolute free energy change. The labels on the data points show the particular structure within U4 in which the insert took place. There is little indication that this relation is anything other than linear (note that the insertion penalty is the result of taking a logarithm of event count ratios). A regression line has a slope of 2.036 bits per kcal/mol and a y-intercept of 2.934 bits. This regression can be used directly to generate priors by calculating the absolute free energy change from the consensus structure and then converting it to insertion penalty prior using the linear regression equation.

The next step is to examine many more ncRNA families to improve the estimated regression. The U4 analysis was done by hand, but a thorough analysis would require automating the process starting from the Rfam multiple alignment and covariance model parameter files. This is a doable, but non-trivial task. The regression estimates also need to be generated for deletions and for helix nucleotide mutations as described in the next section. Unlike insertions in loops, evidence for predictive power of thermodynamic data in helices or for deletions and mutations has not yet been collected.

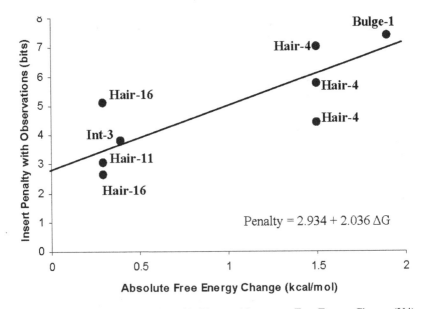

Fig. 6. Regression of Penalties with Observed Inserts on Free Energy Change (U4)

3   Information in Consensus Nucleotides of Helix Neighbors

The emission probabilities for MP states in P nodes are determined by a mixture of observed base pairs in the known family member sequences and an emission prior for MP states. The current Infernal implementation uses a nine-component Dirichlet mixture prior. Basically, there is a set of nine commonly observed probability distributions over the sixteen possible nucleotide pairs estimated from the overall Rfam database. These distributions are similar to eigenvectors in the space of estimated distributions and there is an associated mixture (weighting) coefficient that is similar to an eigenvalue. When the observed distribution is similar to one of the nine components, it reinforces that component at the expense of the others. One of the components with a large mixture coefficient has large probabilities for the Watson-Crick base pairs (AU, UA, CG, and GC) and small probabilities for the others. Another component with large mixture coefficient has high probability for Watson-Crick and wobble pairs (GU and UG). The complete list is in Table 3 of [4]. If there is no variation in the observed within-family base pair distribution at some consensus position (which is very often the case), then the prior for the MP state is the same as the prior for any other MP state with no observed variation and the same consensus base pair.

The choice of MP state emission prior does not currently take into account the consensus environment of the state. In particular, the consensus identities of base pairs stacked before or after the MP state in a helix have no effect. In the RNA secondary structure prediction literature, stacking energies have a significant role in determining the likelihood of a particular helix being formed [18]. It seems that information about whether a particular base-pair substitution is likely to destabilize the rest of the helix would be helpful in determining how strongly the substitution should be penalized in database search.

Table 8. Free energy formation increments for helix propagation (kcal/mol) [14]

Helix segment [a]	Formation free energy	Helix segment	Formation free energy
5'-AA-3' 3'-UU-5'	-0.9	5'-CC-3' 3'-GG-5'	-2.9
5'-AU-3' 3'-UA-5'	-0.9	5'-CG-3' 3'-GC-5'	-2.0
5'-UA-3' 3'-AU-5'	-1.1	5'-GC-3' 3'-CG-5'	-3.4
5'-AC-3' 3'-UG-5'	-2.1	5'-CA-3' 3'-GU-5'	-1.8
5'-AG-3' 3'-UC-5'	-1.7	5'-GA-3' 3'-CU-5'	-2.3

a) Due to rotational symmetry only ten of the sixteen helix segments are unique. Starting at the 5'-end in the lower right for six cases gives exactly the same sequence as starting at the upper left 5'-end of six other cases.

Table 8 shows the measured free energies of formation resulting from adding a base pair to the end of an existing helix where this energy depends on the identity of the nucleotides of the neighboring base pair. To get the total formation energy for the helix, one adds up the stacking energies in Table 8 and then adds other terms (such as helix initiation and symmetry terms) which will be ignored here. For instance, the stacking energy associated with a sequence ...AAC...GUU... would be -0.9 plus -2.1, since the energy for ...AA...UU... is -0.9 and for ...AC...GU... is -2.1.

Table 9. Free energy changes for CG to GC, AU, or UA mutations (kcal/mol) dependent on consensus sequence nucleotides of neighboring base pairs

CG neighbors	CG to GC mutation ΔG	CG to AU mutation ΔG	CG to UA mutation ΔG
5'-AcA-3' 3'-UgU-5'	-0.1	2.1	1.9
5'-AcU-3' 3'-UgA-5'	0.0	2.0	2.0
5'-AcC-3' 3'-UgG-5'	-0.1	2.0	1.8
5'-AcG-3' 3'-UgC-5'	-0.5	1.5	1.4
5'-UcA-3' 3'-AgU-5'	0.0	2.1	2.1
5'-UcU-3' 3'-AgA-5'	0.1	2.0	2.2
5'-UcC-3' 3'-AgG-5'	0.0	2.0	2.0
5'-UcG-3' 3'-AgC-5'	-0.4	1.5	1.6
5'-CcA-3' 3'-GgU-5'	0.4	2.0	1.9
5'-CcU-3' 3'-GgA-5'	0.5	1.9	2.0
5'-CcC-3' 3'-GgG-5'	0.4	1.9	1.8
5'-CcG-3' 3'-GgC-5'	0.0	1.4	1.4
5'-GcA-3' 3'-CgU-5'	0.0	2.0	2.0
5'-GcU-3' 3'-CgA-5'	0.1	1.9	2.1
5'-GcC-3' 3'-CgG-5'	0.0	1.9	1.9
5'-GcG-3' 3'-CgC-5'	-0.4	1.4	1.5

Table 9 shows the sixteen possible three-base-pair stacks with a CG central base pair. Three columns of this table show the change in free energy associated with a mutation of the central base pair from CG to GC, AU, or UA. There is a significant increase in free energy when mutating to AU or UA, which is to be expected since CG bonds with three hydrogen bonds and AU with two hydrogen bonds. However, there are also variations in free energy change for CG to GC mutation that are as much as a third that of the CG to AU or UA changes. This variation depends on the consensus neighbor pairs. Regressions of emission scores at sequence positions with observed mutations on free energy change calculated from consensus neighbors may result in relations similar to that of Figure 6 that could be used as basis for MP state emission priors. This work, however, has not yet been undertaken.

4 Conclusion

The use of experimentally determined thermodynamic data for RNA helix and loop formation appears to have potential for creating more informative priors for loop insertions and deletions and for helix base-pair mutations in ncRNA search based on covariance models. In this paper, examination of the U4 ncRNA family has uncovered some evidence that the absolute value of free energy change calculated from consensus structure predicts insertion probabilities better than the non-family observed-frequency priors based on covariance model node and state types. While certainly not conclusive, this evidence indicates that further investigation of free-energy-based priors may be fruitful.

More analysis needs to be undertaken in the form of automated comparisons of loop insertion and deletion penalties and helix base-pair substitution penalties to changes in free energy based on consensus sequence and secondary structure. Such analysis would firm up the conclusions of this paper and give better estimates of the functional relationships between penalties and free energies which can be used to generate priors. Once reliable functional estimates are obtained, large scale testing of the new priors versus the existing priors could determine conclusively the effectiveness of this method.

Acknowledgments. SFS is grateful for support under NIH Grant Number P20 RR016454 from the INBRE Program of the National Center for Research Resources. KCW would like to thank Dr. Herbert Tsang and Andrew Hendriks for their helpful feedback on an earlier draft of this paper.

References

1. Gesteland, R.F., Cech, T.R., Atkins, J.F.: The RNA World, 3rd edn. Cold Spring Harbor Laboratory Press, New York (2006)
2. Rivas, E., Eddy, S.R.: Secondary Structure Alone is Generally Not Statistically Significant for the Detection of Noncoding RNAs. Bioinformatics 6, 583–605 (2000)
3. Burge, C., Karlin, S.: Prediction of Complete Gene Structures in Human Genomic DNA. Journal of Molecular Biology 268, 78–94 (1997)

4. Nawrocki, E.P., Eddy, S.R.: Query-Dependent Banding (QDB) for Faster RNA Similarity Searches. PLoS Computational Biology 3(3), 540–554 (2007)
5. Smith, S.F.: Covariance Searches for ncRNA Gene Finding. In: IEEE Symposium on Computational Intelligence in Bioinformatics and Computational Biology, pp. 320–326 (2006)
6. Eddy, S.R.: Hidden Markov Models. Current Opinion in Structural Biology 6, 361–365 (1996)
7. Eddy, S.R.: Infernal 0.81 User's Guide (2007), http://infernal.janelia.org/
8. Griffiths-Jones, S., Moxon, S., Marshall, M., Khanna, A., Eddy, S., Bateman, A.: Rfam: Annotating Non-coding RNAs in Complete Genomes. Nucleic Acids Research 33, D121–D141 (2005)
9. Eddy, S.R.: The HMMER User's Guide (2003), http://hmmer.janelia.org/
10. Finn, R.D., Mistry, J., Schuster-Böckler, B., Griffiths-Jones, S., Hollich, V., Lassmann, T., Moxon, S., Marshall, M., Khanna, A., Durbin, R., Eddy, S.R., Sonnhammer, E.L.L., Bateman, A.: Pfam: Clans, Web Tools and Services. Nucleic Acids Research 34, D247–D251 (2006)
11. Brown, M., Hughey, R., Krogh, A., Mian, I.S., Sjölander, K., et al.: Using Dirichlet Mixture Priors to Derive Hidden Markov Models for Protein Families. In: Conference on Intelligent Systems for Molecular Biology, pp. 47–55 (1993)
12. Durbin, R., Eddy, S., Krogh, A., Mitchison, G.: Biological Sequence Analysis: Probabilistic Models of Proteins and Nucleic Acids. Cambridge University Press, Cambridge (1998)
13. Sjölander, K., Karplus, K., Brown, M., Hughey, R., Krogh, A., Mian, I., Haussler, D.: Dirichlet Mixtures: A Method for Improving Detection of Weak but Significant Protein Sequence Homology. Comp. Appl. BioSci. 12, 327–345 (1996)
14. Freier, S., Kierzek, R., Jaeger, J., Sugimoto, N., Caruthers, M., Neilson, T., Turner, D.: Improved Free-Energy Parameters for Predictions of RNA Duplex Stability. Proc. Nat. Acad. Sci. USA 83, 9373–9377 (1986)
15. Lobert, P.E., Escriou, N., Ruelle, J., Michiels, T.: A Coding RNA Sequence Acts as a Replication Signal in Cardioviruses. Proc. Nat. Acad. Sci. USA 96, 11560–11565 (1999)
16. Calin, G.A., Dumitru, C.D., Shimizu, M., Bichi, R., Zupo, S., Noch, E., Aldler, H., Rattan, S., Keating, M., Rai, K., Rassenti, L., Kipps, T., Negrini, M., Bullrich, F., Croce, C.M.: Frequent Deletions and Down-Regulation of Micro-RNA Genes miR15 and miR16 at 13q14 in Chronic Lymphocytic Leukemia. Proc. Nat. Acad. Sci. USA 99, 15524–15529 (2002)
17. Zucker, M.: Computer Prediction of RNA Structure. Methods in Enzymology 180, 262–288 (1989)
18. Wiese, K.C., Deschênes, A.A., Hendriks, A.G.: RnaPredict–An Evolutionary Algorithm for RNA Secondary Structure Prediction. IEEE/ACM Transactions on Computational Biology and Bioinformatics 5, 25–41 (2008)
19. Raghunathan, P.L., Guthrie, C.: A Spliceosomal Recycling Factor that Reanneals U4 and U6 Small Nuclear Ribonucleoprotein Particles. Science 279, 857–860 (1998)
20. Smith, S.F., Wiese, K.C.: Improved Covariance Model Parameter Estimation Using RNA Thermodynamic Properties. In: International Conference on Bio-Inspired Models of Network, Information, and Computing Systems - Bionetics (2007)

An Evolutionary Approach to the Non-unique Oligonucleotide Probe Selection Problem

Lili Wang, Alioune Ngom*, Robin Gras, and Luis Rueda**

School of Computer Science, 5115 Lambton Tower
University of Windsor, 401 Sunset Avenue
Windsor, Ontario, N9B 3P4, Canada
{wang111v, angom, rgras, lrueda}@uwindsor.ca

Abstract. In order to accurately measure the gene expression levels in microarray experiments, it is crucial to design *unique*, highly specific and sensitive oligonucleotide probes for the identification of biological agents such as genes in a sample. Unique probes are difficult to obtain for closely related genes such as the known strains of HIV genes. The *non-unique* probe selection problem is to select a probe set that is able to uniquely identify targets in a biological sample, while containing a minimal number of probes. This is an NP-hard problem. We define a probe selection function that allows to decide which are the best probes to include in or exclude from a candidate probe set. We then propose a new deterministic greedy heuristic that uses the selection for solving the non-unique probe selection problem. Finally, we combine the selection function with an evolutionary method for finding near minimal non-unique probe sets. When used on benchmark data sets, our greedy method outperforms current greedy heuristics for non-unique probe selection in most instances. Our genetic algorithm also produced excellent results when compared to advanced methods introduced in the literature for the non-unique probe selection problem.

Keywords: Microarrays, Probe Selection, Targets, Hybridization, Separation and Coverage, Heuristics, Genetic Algorithms, Optimization, Dominated Row.

1 Introduction

Oligonucleotide microarrays, commonly known as gene chips, are widely used techniques in molecular biology, providing a fast and cost-effective method for monitoring the expression of thousands of genes simultaneously [2][8]. For this, short strands of known DNA sequences, called oligonucleotide *probes*, are attached to specific positions on a chip's surface. The length of a probe is typically between 8 to 25 base-pairs. Fluorescently labeled cDNA molecules of RNA samples are then

* Research partially funded by Canadian NSERC Grant #RGPIN228117-2006 and CFI grant #9263.
** Research partially funded by Chilean FONDECYT Grant #1060904.

C. Priami et al. (Eds.): Trans. on Comput. Syst. Biol. X, LNBI 5410, pp. 143–162, 2008.
© Springer-Verlag Berlin Heidelberg 2008

probed over this surface. Some of these cDNAs will bind to complementary probes by means of *hybridization*. The amount of cDNA hybridized to each position on the microarray can be inferred from fluorescence measurements [2].

In order to measure the expression level of a specific gene in a sample, one must design a microarray containing probes that are complementary to gene segments called *targets*. Typically, the total length of a probe used to hybridize a gene is only a small fraction of the length of the gene [8]. The success of a microarray experiment depends on the quality of the probes: that is, how well each probe hybridizes to its target. Expression levels can only be accurately measured if each probe hybridizes to its target only, given the target is present in the biological sample at any concentration. However, choosing good probes is a difficult task since different sequences have different hybridization characteristics.

A probe is *unique*, if it is designed to hybridize to a single target. However, there is no guarantee that unique probes will hybridize to their intended targets only. Cross-hybridization, (hybridization to non-target sequences), self-hybridization (a probe hybridizing to itself) and non-sensitive hybridization (a probe may not hybridize to its present target) are hybridization errors that usually occur and must be taken into consideration for accurate measurement of gene expression levels. Many parameters such as secondary structure, salt concentration, GC content, free energy and melting temperature also affect the hybridization quality of probes [8], and their values must be carefully determined to design high quality probes.

It is particularly difficult to design unique probes for closely related genes that are to be identified. Too many targets can be similar and hence cross-hybridization errors increase substantially. An alternative approach is to devise a method that can make use of *non-unique* probes, i.e. probes that are designed to hybridize to at least one target [4]. The use of non-unique probes in oligonuleotide microarray experiments has many advantages. First, a much smaller probe set can be used with non-unique probes than can be with unique probes; unique probes hybridize to unique targets only, and hence the probe set cannot be minimized. Second, minimizing the number of probes in an experiment is a very reasonable objective, since it is proportional to the cost of the experiment; one cannot reduce this cost with unique probes. Third, the use of non-unique probes allows the possibility to *pack* more DNA molecules on a chip surface, thus increasing the number of hybridization experiments that can be performed on the chip.

The *non-unique probe selection problem* is to determine the smallest possible set of probes that is able to identify all targets present in a biological sample. This is an NP-hard problem [3] and several approaches to its solution have been proposed recently [3][4][5][6][7][9][10] (see Section 3). In this paper, we introduce a non-random greedy heuristic and a genetic algorithm to find near minimal non-unique probe sets. Our methods are based on the definition of a probe selection function which helps to decide which is the best probe to include in or exclude from a candidate solution.

The remainder of the paper is organized as follows. In Section 2, we formally define the problem and formulate it as an integer linear programming problem. In Section 3, we discuss previous work on this problem. In Section 5, we introduce the probe selection function and describe the deterministic greedy heuristic which uses the selection function. In Section 6, we devise a genetic algorithm based on the selection function introduced in Section 4. Experimental results are discussed in Section 7 and we conclude with a summary of our work in Section 8.

2   Non-unique Probe Selection Problem

Given a target set, $T = \{t_1, \ldots, t_m\}$, and probe set, $P = \{p_1, \ldots, p_n\}$, an $m \times n$ *target-probe incidence matrix* $H = [h_{ij}]$ is such that $h_{ij} = 1$, if probe p_j hybridizes to target t_i, and $h_{ij} = 0$ otherwise. Table 1 shows an example of a matrix with $m = 4$ targets and $n = 6$ probes. A probe p_j *separates* two targets, t_i and t_k, if it is a substring of either t_i or t_k, that is, if $|h_{ij} - h_{kj}| = 1$. For example, if $t_i =$ AGGCAATT and $t_k =$ CCATATTGG, then probe $p_j =$ GCAA separates t_i and t_k, since it is a substring of t_i only, whereas probe $p_l =$ ATT does not separate t_i and t_k, since it is a substring of both targets [5]. Two targets, t_i and t_k, are *s-separated*, $s \geq 1$, if there exist at least s probes such that each separates t_i and t_k; in other words, the Hamming distance between rows i and k in H is at least s. For example, in Table 1 targets t_2 and t_4 are 4-separated. A target t is *c-covered*, $c \geq 1$, if there exist at least c probes such that each hybridizes to t. In Table 1, target t_2 is 3-covered. Due to hybridization errors in microarray experiments, it is required that any two targets be s_{min}-separated and any target be c_{min}-covered; usually, we have $s_{min} \geq 2$ and $c_{min} \geq 2$. These two requirements are called *separation constraints* and *coverage constraints*.

Given a matrix H, the objective of the non-unique probe selection problem is to find a minimal subset of probes that determines the presence or absence of specified targets, and such that all constraints are satisfied. In Table 1, if $s_{min} = c_{min} = 1$ and assuming that exactly one of $t_1, \ldots, t_4$ is in the sample, then the goal is to select a minimal set of probes that allows us to infer the presence or absence of a single target. In this case, a minimal solution is $\{p_1, p_2, p_3\}$ since for target t_1, probes p_1 and p_2 hybridize while p_3 does not; for target t_2, probes p_1 and p_3 hybridize while p_2 does not; for target t_3, probes p_2 and p_3 hybridize while p_1 does not; and finally for target t_4, only probe p_3 hybridizes. Thus, each

Table 1. Example of a target-probe incidence matrix

	p_1	p_2	p_3	p_4	p_5	p_6
t_1	1	1	0	1	0	1
t_2	1	0	1	0	0	1
t_3	0	1	1	1	1	1
t_4	0	0	1	1	1	0

single target will be identified by the set $\{p_1, p_2, p_3\}$, if it is the only target present in the sample; moreover, all constraints are satisfied. For $s_{\min} = c_{\min} = 2$, a minimal solution that satisfies all constraints is $\{p_2, p_3, p_5, p_6\}$. Of course, the set $\{p_1, \ldots, p_6\}$ is a solution but it is not minimal, and therefore is not cost-effective.

Our prior assumption above, that only a single target is present in the sample, creates some difficulties when in reality multiple targets are in the sample. Let $s_{\min} = c_{\min} = 1$, then selecting $\{p_1, p_2, p_3\}$ as before, results in the hybridization of all probes when both targets t_1 and t_2 are in the sample. Hence, we cannot distinguish between the case where pair (t_1, t_2) is in the sample and where t_3 is also in the sample. Using $\{p_2, p_3, p_5\}$, however, instead of $\{p_1, \ldots, p_6\}$ will resolve this experiment. Let $s_{\min} = c_{\min} = 2$, then any of pairs (t_1, t_3), (t_1, t_4), (t_2, t_3), and (t_3, t_4) will cause the hybridization of all probes of $\{p_2, p_3, p_5, p_6\}$; thus, we can determine the presence of target t_3 for instance, but cannot determine which other target is present.

The solution to the multiple target problem is to use *aggregated targets* that represent the presence or absence of *subsets* of targets in the samples [3]. An aggregated target t^a is thus a subset of targets, and the single target separation problem is a special case of the multiple target separation problem, i.e. such that each aggregated target is a singleton set of a single target.

Stated formally, given an $m \times n$ target-probe incidence matrix H with a target set $T = \{t_1, \ldots, t_m\}$ and a probe set $P = \{p_1, \ldots, p_n\}$, and a minimum coverage parameter $c_{\min}$, a minimum separation parameter $s_{\min}$ and a parameter $d_{\max} \geq 1$, the aim of the non-unique probe selection problem is to determine a subset $P_{\min} = \{q_1, q_2, \cdots, q_s\} \subseteq P$ of probes such that:

1. $s = |P_{\min}| \leq n$ is minimal.
2. Each target $t \in T$ is $c_{\min}$-covered by some probes in $P_{\min}$.
3. Each pair of aggregated targets $(t_x^a, t_y^a) \in 2^T \times 2^T$, with $|t_x^a|, |t_y^a| \leq d_{\max}$, is $s_{\min}$-separated by some probes in $P_{\min}$.

In Point 3. above, we seek to guarantee $s_{\min}$-separation not only between pairs of single targets but also between pairs of *small* subsets of targets $t_x^a, t_y^a \in 2^T$ (2^T is the power set of T), each subset up to cardinality $d_{\max}$. These requirements are called the *group separation constraints* [3]. Given a subset of targets t^a, we denote by w^{t^a} the vector that results from applying the logical OR to the rows of H associated with targets from t^a. This vector, called the *signature* of t^a, represents the set of all probes hybridizing to t^a as a binary vector and is defined by $w_j^{t^a} = \max_{i \in t^a} h_{ij}, 1 \leq j \leq n$. Note that aggregated targets t_x^a and t_y^a are $s_{\min}$-separated, if and only if the Hamming distance between $w^{t_x^a}$ and $w^{t_y^a}$ is at least $s_{\min}$.

This problem has a polynomial number of constraints for the single target separation, and at least an exponential number of constraints for the aggregated target separation. In [3], it was proved to be NP-hard by performing a reduction from the *set covering* problem. This problem is NP-hard even for single targets as well as for $c_{\min} = 1$ or $s_{\min} = 1$.

It is not trivial to compute the target-probe incidence matrix H at the beginning. The number of potential non-unique probes is very large and only a small fraction of these probes satisfy the usual criteria of *specificity* (probes should not cross-hybridize or be self-complementary), *sensitivity* (probes should hybridize to their low-abundant targets) and *homogeneity* (probes should exhibit the same hybridization affinity, that is, the same free energy at a given temperature and salt concentration) required for good oligonucleotide probes [8]. The aim of this paper is not to compute the target-probe incidence matrix H but to find the optimal non-unique probe set given H.

The work of [4] was the first to formulate the non-unique probe selection problem for single targets as an *integer linear programming* (ILP) problem. Let $C = \{(i,k) \mid 1 \leq i < k \leq m\}$ be the set of all combinations of target indices. Assign $x_j = 1$ if probe p_j is chosen and 0 otherwise. Then, we have the following ILP formulation:

$$\text{Minimize: } \sum_{j=1}^{n} x_j \ . \tag{1}$$

Subject to:

$$x_j \in \{0,1\} \qquad 1 \leq j \leq n \ , \tag{2}$$

$$\sum_{j=1}^{n} h_{ij} x_j \geq c_{\min} \qquad 1 \leq i \leq m \ , \tag{3}$$

$$\sum_{j=1}^{n} |h_{ij} - h_{kj}| x_j \geq s_{\min} \qquad 1 \leq i < k \leq m \ . \tag{4}$$

Function (1) minimizes the number of probes. The probe selection variables are binary-valued in Restriction (2). Constraints (3) and (4) are the coverage and separation constraints, respectively. Note that Constraints (4) are for single targets only. As opposed to this, in [3], the following ILP formulation was proposed, which also includes the group separation constraints for aggregated targets:

$$\text{Minimize: } \sum_{j=1}^{n} x_j \ . \tag{5}$$

Subject to:

$$x_j \in \{0,1\} \qquad 1 \leq j \leq n \ , \tag{6}$$

$$\sum_{j=1}^{n} \left| \omega_j^{t_x^{a}} - \omega_j^{t_y^{a}} \right| x_j \geq \min \left\{ d, \sum_{j=1}^{n} \left| \omega_j^{t_x^{a}} - \omega_j^{t_y^{a}} \right| \right\} \qquad \forall (t_x^{a}, t_y^{a}) \in 2^T \times 2^T \ , \tag{7}$$

$$|t_x^{a}|, |t_y^{a}| \leq d_{\max} \ ,$$
$$t_x^{a} \neq t_y^{a} \ .$$

where $c_{\min} = s_{\min} = d$. Here, Constraints (7) are the group separation constraints, which also include the single target separation constraints. The coverage constraints are also satisfied by (7) with $t_x^{a} = \emptyset$ and $t_y^{a} = \{t_i\}$ for $1 \leq i \leq m$.

In this paper, we solve the original ILP formulation of [4] using a deterministic greedy heuristic and a genetic algorithm. Thus, we are not concerned with aggregated target separations. Note that one can easily check if the original set of candidate probes satisfy all the constraints. If not, then there are no feasible solutions. In this case, we can insert *unique virtual probes* in the original probe set only for those targets or target-pairs that are not $c_{\min}$-covered or $s_{\min}$-separated. This will ensure the existence of feasible solutions.

3 Previous Work

Schliep *et al.* [7] first introduced the non-unique probe selection problem and described a simple but fast greedy heuristic which computes an approximate solution that guarantees $s_{\min}$-separation for pairs of small aggregated targets. Klau *et al.* [4] proposed the first ILP formulation for this problem but for single targets only. They first applied the greedy heuristic of [7] to reduce the initial candidate probe set and then used an ILP solver such as CPLEX software to further reduce the result of applying [7]'s heuristic. Their ILP solutions outperformed those of [7] in all instances. Meneses *et al.* [5] proposed a greedy non-random heuristic for single targets only. They greatly outperformed the ILP method of [4] in the largest instance in the paper but performed poorly in all the other (much smaller) instances. They first used local search and sorting to construct a probe set with a minimal number of violated constraints, and then further reduce this set by iteratively removing probes in such a way that a feasible solution is maintained. Klau *et al.* [3] extended the ILP formulation of [4] to include aggregated targets and prescribed a *branch-and-cut* heuristic for solving such ILP problem. They also proved that the non-unique probe selection problem is NP-hard. Ragle *et al.* [6] developed an *optimal cutting-plane* ILP heuristic, for single targets only, to find optimal solutions within practical computational limits. Their method constitutes a *branch-and-bound* approach that relaxes a large constraint set in order to find and improve the lower bound on the number of probes required in an optimal solution, until an optimal solution is obtained. They greatly outperformed [3] and [4] in 8 out of 11 instances including the largest instance (of size 679 targets × 15139 probes) which they reduced to 1962 probes. Wang *et al.* [10] modified the heuristic of [5] to use a selection function defined over a candidate probe set (see Section 4). The selection function allowed to decide which probes are the best to be included in a solution, based on their ability to help satisfy the constraints. The approach of [10] outperformed that of [5] in 9 out of 13 instances including the largest instance which was reduced from 15139 probes to 2084 probes (even better than in [3] and [4] where that same large instance is reduced to 3158 probes). Recently, Wang *et al.* [9] combined the selection function of [10] with a genetic algorithm and produced results that are at least comparable to (and in most cases, better than) those obtained by the method of [6].

4   Probe Selection Function

In this section, we discuss the probe selection function defined previously in [10]. The selection function allows to decide which probes are the best to be included in a candidate solution according to their ability to help satisfy the constraints. In Section 5, we modify the heuristic of [5] to use a probe selection function defined over a candidate probe set.

We want to select a minimum number of probes such that each target is $c_{\min}$-covered and each target-pair is $s_{\min}$-separated. To answer the question of which probe to select, we need to discuss some properties of a minimal solution, $P_{\min}$. Consider a target-probe incidence matrix, H, the parameters $c_{\min}$ and $s_{\min}$, the initial feasible candidate probe set, $P = \{p_1, \ldots, p_n\}$, and the target set $T = \{t_1, \ldots, t_m\}$. Let P_{t_i} be the set of probes hybridizing to target t_i, and $P_{t_{ik}}$ be the set of probes separating the target-pair t_{ik}. Since there are m coverage constraints (i.e., number of targets) and $\frac{m(m-1)}{2}$ separation constraints (i.e., number of target-pairs), $P_{\min}$ is the union of $m + \frac{m(m-1)}{2}$ subsets of size $c_{\min}$ or $s_{\min}$, each $c_{\min}$-covering a target or $s_{\min}$-separating a target-pair, respectively. That is:

$$P_{\min} = \left\{ \bigcup_{1 \leq i \leq m} P_i \right\} \cup \left\{ \bigcup_{1 \leq i < k \leq m} P_{ik} \right\}, \tag{8}$$

where $P_i \subseteq P_{t_i}$ and $P_{ik} \subseteq P_{t_{ik}}$ are, respectively, the $c_{\min}$-subsets and the $s_{\min}$-subsets (an l-subset is a subset of cardinality l) selected for a minimal solution $P_{\min}$. Equation (8) also shows that $P_{\min}$ can be constructed by determining all its P_i and P_{ik} subsets such that their union is minimal.

A $c_{\min}$-subset $P_i \subseteq P_{t_i}$ (correspondingly, a $s_{\min}$-subset $P_{ik} \subseteq P_{t_{ik}}$) is an *essential covering subset* (correspondingly, *essential separating subset*), if and only if $P_i = P_{t_i}$ (correspondingly, $P_{ik} = P_{t_{ik}}$). In other words, there are only $c_{\min}$ probes that hybridize to t_i and only $s_{\min}$ probes that separate t_{ik}. In Table 1, for instance, subsets $P_2 = \{p_1, p_3, p_6\} = P_{t_2}$ and $P_4 = \{p_3, p_4, p_5\} = P_{t_4}$ are essential covering subsets if $c_{\min} = 3$. Probes belonging to essential (covering or separating) subsets are called *essential probes*. Essential probes or subsets, if they exist, must be contained in any feasible solution; that is, removing any such probe will make the solution unfeasible. A *redundant probe* is the one for which a feasible solution remains feasible when the probe is removed; such probes are discarded and hence are contained in no feasible solution. Note that a probe may be redundant for some candidate solutions but non-redundant for others. Our problem is then to find the *non-essential non-redundant probes* for each $P_i \subseteq P_{t_i}$ and $P_{ik} \subseteq P_{t_{ik}}$ to obtain a minimal solution $P_{\min}$.

Subsets P_{t_i} and $P_{t_{ik}}$ have an interesting property that will be used to decide which probes to select for $P_{\min}$. We have $c_{\min} \leq |P_{t_i}| \leq n$ and $s_{\min} \leq |P_{t_{ik}}| \leq n$. Subsets P_{t_i} of size $|P_{t_i}| = c_{\min}$ contain only essential probes; such probes contribute in all feasible solutions. Subsets P_{t_i} of size $|P_{t_i}| = n$ contain many redundant probes; such probes do not contribute in any minimal solution. Subsets P_{t_i} of size $c_{\min} < |P_{t_i}| < n$ contain non-essential non-redundant probes with varying degrees of contribution to feasible solutions; smaller subsets contribute its probes to more feasible (minimal) solutions than larger subsets. The same is true for subsets $P_{t_{ik}}$.

We want to choose a minimum number of probes such that all constraints are satisfied. Given the property of the subsets P_{t_i} and $P_{t_{ik}}$ discussed above, one can associate with each probe p a *degree of contribution* $D(p)$ to minimal solutions, in order to help decide which probe to consider the best for inclusion in a feasible candidate solution. Probes that appear in the largest number of minimal sets are preferred over those that appear in fewer minimal sets. Thus, *good probes* are those with the highest degree of contribution to minimal solutions (the best being the essential probes), whereas *bad probes* are those with the lowest degree of contribution to minimal solutions (the worst being redundant probes). The following three sections define the probe selection function over the candidate probe set, which is used in Sections 5 and 6. Below, we assume that the initial candidate probe set is feasible. If not, we insert into P a sufficient number of unique virtual probes for each target t_i or each target-pair t_{ik} in which a constraint is not satisfied; that is, $(c_{\min} - |P_{t_i}|)$ and $(s_{\min} - |P_{t_{ik}}|)$ virtual probes are included.

4.1   Coverage Function

We want to choose the minimum number of probes such that each target is $c_{\min}$-covered. Given matrix H, parameter $c_{\min}$, candidate probe set $P = \{p_1, \ldots, p_n\}$ and target set $T = \{t_1, \ldots, t_m\}$, we defined the function cov $: P \times T \rightarrow [0, 1]$ in [10] as follows:

$$\mathrm{cov}(p_j, t_i) = h_{ij} \times \frac{c_{\min}}{|P_{t_i}|}, \quad p_j \in P_{t_i}, \quad t_i \in T , \tag{9}$$

where $0 \leq \mathrm{cov}(p_j, t_i) \leq 1$ and P_{t_i} is the set of probes hybridizing to target t_i; $\mathrm{cov}(p_j, t_i)$ is the amount that p_j contributes to satisfy the coverage constraint for target t_i. For target t_i, p_j is likely to be redundant for a larger value of $|P_{t_i}|$ and likely to be necessary for a smaller value of $|P_{t_i}|$. We defined the *coverage function* $C : P \rightarrow [0, 1]$ in [10] as follows:

$$C(p_j) = \max_{t_i \in T_{p_j}} \{\mathrm{cov}(p_j, t_i) \mid 1 \leq j \leq n\} , \tag{10}$$

where T_{p_j} is the set of targets covered by p_j. $C(p_j)$ is the maximum amount that p_j can contribute to satisfy the minimum coverage constraints. Table 2 shows the coverage function table produced from Table 1.

Function C favors the selection of probes that $c_{\min}$-cover targets t_i that have the smallest subsets P_{t_i}; these are the essential or near-essential covering probes. In Table 2 for example, target t_2 has the minimal value $|P_{t_2}| = 3$, and hence any probe that covers it can be selected first. As discussed earlier, such probes contribute to most minimal solutions and therefore must be included in a candidate solution. In particular, function C guarantees the selection of near-essential covering probes that $c_{\min}$-cover *dominated targets*; t_i

Table 2. Example of a coverage function table

	p_1	p_2	p_3	p_4	p_5	p_6
t_1	$\frac{c_{\min}}{4}$	$\frac{c_{\min}}{4}$	0	$\frac{c_{\min}}{4}$	0	$\frac{c_{\min}}{4}$
t_2	$\frac{c_{\min}}{3}$	0	$\frac{c_{\min}}{3}$	0	0	$\frac{c_{\min}}{3}$
t_3	0	$\frac{c_{\min}}{5}$	$\frac{c_{\min}}{5}$	$\frac{c_{\min}}{5}$	$\frac{c_{\min}}{5}$	$\frac{c_{\min}}{5}$
t_4	0	0	$\frac{c_{\min}}{3}$	$\frac{c_{\min}}{3}$	$\frac{c_{\min}}{3}$	0
C	$\frac{c_{\min}}{3}$	$\frac{c_{\min}}{4}$	$\frac{c_{\min}}{3}$	$\frac{c_{\min}}{3}$	$\frac{c_{\min}}{3}$	$\frac{c_{\min}}{3}$

dominates t_k if $P_{t_k} \subset P_{t_i}$. In Table 2, for example, t_3 dominates t_4 since $P_{t_4} = \{p_3, p_4, p_5\} \subset \{p_2, p_3, p_4, p_5, p_6\} = P_{t_3}$. Any $c_{\min}$-cover of the dominated target t_k will also $c_{\min}$-cover all its dominant targets, and therefore, more targets are $c_{\min}$-covered. Probes from the dominated target t_k have larger cov values than probes from all its dominant targets since $|P_{t_k}| < |P_{t_i}|$ and hence they will be selected first.

4.2 Separation Function

We want to choose the minimum number of probes such that each target-pair is $s_{\min}$-separated. We defined the function sep : $P \times T^2 \rightarrow [0, 1]$ in [10] as follows:

$$\text{sep}(p_j, t_{ik}) = |h_{ij} - h_{kj}| \times \frac{s_{\min}}{|P_{t_{ik}}|}, \quad p_j \in P_{t_{ik}}, \quad t_{ik} \in T^2 , \qquad (11)$$

where $0 \leq \text{sep}(p_j, t_{ik}) \leq 1$ and $P_{t_{ik}}$ is the set of probes separating target-pair t_{ik}; $\text{sep}(p_j, t_{ik})$ is what p_j can contribute to satisfy the separation constraint for target-pair t_{ik}. We defined the *separation function* $S : P \rightarrow [0, 1]$ in [10] as follows:

$$S(p_j) = \max_{t_{ik} \in T^2_{p_j}} \{\text{sep}(p_j, t_{ik}) \mid 1 \leq j \leq n\} , \qquad (12)$$

where $T^2_{p_j}$ is the set of target-pairs separated by p_j. $S(p_j)$ is the maximum amount that p_j can contribute to satisfy the minimum separation constraints. Table 3 shows the separation function table produced from Table 1.

Function S favors the selection of probes that $s_{\min}$-separate target-pairs t_{ik} that have the smallest subsets $P_{t_{ik}}$; these are the essential or near-essential separating probes. In particular, function S favors the selection of near-essential separating probes that $s_{\min}$-separate *dominated target pairs*.

4.3 Selection Function

We want to select the minimum number of probes such that all coverage and separation constraints are satisfied; that is, we must select a probe according to its ability to help satisfy both coverage *and* separation constraints. In [10], we combined Functions (10) and (12) into a single probe selection function $D : P \rightarrow [0, 1]$ as follows:

$$D(p_j) = \max\{(C(p_j), S(p_j)) \mid 1 \leq j \leq n\} . \qquad (13)$$

Table 3. Example of a separation function table

	p_1	p_2	p_3	p_4	p_5	p_6
t_{12}	0	$\frac{s_{min}}{3}$	$\frac{s_{min}}{3}$	$\frac{s_{min}}{3}$	0	0
t_{13}	$\frac{s_{min}}{3}$	0	$\frac{s_{min}}{3}$	0	$\frac{s_{min}}{3}$	0
t_{14}	$\frac{s_{min}}{5}$	$\frac{s_{min}}{5}$	$\frac{s_{min}}{5}$	0	$\frac{s_{min}}{5}$	$\frac{s_{min}}{5}$
t_{23}	$\frac{s_{min}}{4}$	$\frac{s_{min}}{4}$	0	$\frac{s_{min}}{4}$	$\frac{s_{min}}{4}$	0
t_{24}	$\frac{s_{min}}{4}$	0	0	$\frac{s_{min}}{4}$	$\frac{s_{min}}{4}$	$\frac{s_{min}}{4}$
t_{34}	0	$\frac{s_{min}}{2}$	0	0	0	$\frac{s_{min}}{2}$
S	$\frac{s_{min}}{3}$	$\frac{s_{min}}{2}$	$\frac{s_{min}}{3}$	$\frac{s_{min}}{3}$	$\frac{s_{min}}{3}$	$\frac{s_{min}}{2}$

$D(p_j)$ is the degree of contribution of p_j, that is, the maximum amount required for p_j to satisfy all constraints. D ensures that all essential probes p_j will be selected for inclusion in the subsequent candidate solution, since $C(p_j) = 1$ or $S(p_j) = 1$. With our definition of D, probes p that cover dominated targets or separate dominated target-pairs have the highest $D(p)$ values. By selecting a probe p to cover a dominated target t_i or to separate a dominated target-pair t_{ik}, we are also selecting p to cover as many targets as possible (all targets that dominate t_i) or to separate as many target-pairs as possible (all target-pairs that dominate t_{ik}). This is the main greedy strategy inherent to our heuristic introduced in the next section.

5   Deterministic Greedy Heuristic for Non-unique Probes

In this section, we propose a deterministic greedy heuristic that filters out *bad probes* as in Meneses *et al.* [5]. That work used no selection function to decide which probes to filter out; probes are removed as long as the feasibility of the given candidate solution is not compromised: these are the *bad probes* in [5]. Also the approach of [5] used no random selection at any time in the algorithm. They initially sort the probes in increasing order of the number of targets they hybridize and then select probes, in this order, for inclusion in a candidate solution. The authors then scan the resulting feasible candidate probe set to test each probe for possible redundancy and remove any redundant probe. No additional information is used to direct the search. In the data sets, the range of the number of targets to which each probe hybridize is very small (for example, that range is [1, 40] for the largest data set in [5]) and many probes hybridize to the same number of targets. Thus, given two candidate probes, it is not easy to identify which probe is better than another for inclusion into a candidate solution. In our heuristic, the $D(p)$ values store much more information about the current probe set, in such a way that the algorithm can decide which probes to consider best for selection.

The *Dominated Row Covering* (DRC) heuristic described in [10] *decides* at any given moment which probe is the best to include in or to exclude from a candidate solution. The decision to include a probe is made *locally*, that is, it depends only on the current probe set to select from. Likewise, the decision to

1. Initialization
   (a) Compute $D(p)$ for all $p \in P$ using Equations (9)–(13).
   (b) $P_{\text{ini}} = \{p \in P \mid D(p) = 1\}$ /* essential probes in initial solution */.
2. Construction
   (a) $P_{\text{sol}} = P_{\text{ini}}$
   (b) Sort $P \setminus P_{\text{sol}}$ in decreasing order of $D(p)$.
   (c) For each target t_i not c_{min}-covered by P_{sol}.
       i. n_i = number of missing probes required to c_{min}-cover t_i.
       ii. $P_{\text{sol}} = P_{\text{sol}} \cup \bigcup_{l=1}^{l=n_i} \{$next highest-degree $p_l \in P \setminus P_{\text{sol}}$ that covers $t_i\}$.
   (d) For each target-pair t_{ik} not s_{min}-separated by P_{sol}
       i. n_{ik} = number of missing probes required to s_{min}-separate t_{ik}.
       ii. $P_{\text{sol}} = P_{\text{sol}} \cup \bigcup_{l=1}^{l=n_{ik}} \{$next highest-degree $p_l \in P \setminus P_{\text{sol}}$ that separates $t_{ik}\}$.
3. Reduction
   (a) $P_{\text{min}} = P_{\text{sol}}$.
   (b) $H = H|_{P_{\text{min}}}$, /* the restriction of matrix H to the probes in P_{min} */.
   (c) Compute $D(p)$ for all $p \in P_{\text{min}}$.
   (d) Sort $P_{\text{del}} = \{p \in P_{\text{min}} \mid D(p) < 1\}$ in increasing order of $D(p)$.
   (e) If $P_{\text{min}} \setminus \{p\}$ is feasible for each $p \in P_{\text{del}}$ then
          $P_{\text{min}} = P_{\text{min}} \setminus \{p\}$.
   (f) Return final P_{min}.

Fig. 1. Dominated Row Covering Heuristic

exclude a probe depends only on the current candidate probe set. No past or global information to find a near minimal probe set is used. The heuristic, shown in Figure 1 consists of three phases: *initialization, construction,* and *reduction.*

In the *Initialization phase,* we compute the initial $D(p)$ value for each probe $p \in P$ given matrix H, and create an initial and possibly non-feasible solution P_{ini}, which contains essential probes only. In the *Construction phase,* we repeatedly insert high-degree probes into P_{ini} until an initial feasible solution P_{sol} is obtained. In the *Reduction phase,* we reduce P_{sol} by repeatedly removing low-degree probes so as to obtain a final near minimal feasible solution P_{min}.

DRC is *greedy* as it performs a non-random hill-climbing search in the space 2^P (the power set of P) by selecting only highest-degree probes to satisfy the constraints and by removing only lowest-degree probes to maintain the feasibility of a candidate solution. Solution P_{min}, therefore, is not necessarily optimal. In DRC, probes p that cover dominated targets (that is, dominated rows of the coverage matrix, e.g., Table 2) or that separate dominated target-pairs (that is, cover dominated rows of the separation matrix, e.g., Table 3) have higher $D(p)$ values than probes that cover dominant targets or separate dominant target-pairs, and hence they are always selected first.

Note on the Computational Complexity of DRC Heuristic. In the worst case, the time complexity of the DRC is dominated by the separation constraint satisfactions in the *Construction Phase.* There are potentially $\frac{m^2 - m}{2}$ target-pairs that are not yet s_{min}-separated by P_{sol}. Testing a target-pair t_{ik} for s_{min}-separation takes at most n steps (value $n_{ik} \leq s_{\text{min}} \ll n$ is returned after this test). Searching the highest-degree probe p_l, for each missing probe, to include into P_{sol} takes at most n steps. Therefore DRC runs in $O(n^2 m^2)$ time. In practice, DRC runs much faster than $O(n^2 m^2)$ because $n_{ik} \leq s_{\min} < c_{\min} \lll m \leq n$.

6   Genetic Algorithm for Non-unique Probes

Beasley *et al.* [1] presented a genetic algorithm-based heuristic for non-unicost set-covering problems in which they proposed modifications to the basic simple genetic algorithm (GA). We adapted, with some modifications, the approach of [1] to the non-unique probe selection problem, which is itself a set covering problem. Our GA makes use of the probe selection function defined in Section 4 . The next sections describe our genetic approach.

6.1   Representation and Fitness Function

The first step in designing a genetic algorithm is to devise a suitable representation scheme. Given the initial candidate probe set $P = \{p_1, \ldots, p_n\}$, we want to find a feasible subset $P_{\min} \subseteq P$ of minimal cardinality. Therefore, the search space is the power set of P, denoted by 2^P; that is the set of all subsets of P. The usual binary representation for subsets $S \subseteq P$ is thus an obvious choice for the non-unique probe selection problem. Given a solution $S \subseteq P$ we use an n-bit binary string $s = s_1 \ldots s_n$ as our chromosome representation where, $s_j = 1$ if $p_j \in S$ and $s_j = 0$ if $p \notin S$.

The fitness of an individual s is directly related to its objective value, which corresponds to the number of probes in its associated subset S. That is, the fitness of s is defined as

$$f(s) = \sum_{j=1}^{n} s_j$$

6.2   Parent Selection Operator

When a population converges, the range of fitness values in the population reduces. In order to continuously favor selecting the fittest individuals, we use tournament selection and fitness scaling. Fitness scaling maps a raw fitness value onto a new value by subtracting a suitable value from the raw fitness, for example

$$f_c^s = f_c - \min_{1 \leq c \leq N} f_c$$

where f_c and f_c^s denote the raw fitness and the scaled fitness of individual c, and N is the population size.

6.3   Crossover Operator

We use the *fusion operator* of [1], a *generalized fitness-based* crossover which considers both the structure and the relative fitness of each parent solution, and produces just a single child. This crossover focuses on the differences in the parents and hence is more capable of generating new solutions when the parents are similar. Also, the fittest parent is likely to contribute more to the child's fitness. Let $f^s_{P_1}$ and $f^s_{P_2}$ be the scaled fitness values of the parents P_1 and P_2 respectively, and let C denote the child solution, then for $1 \le j \le n$ we have

1. if $P_{1j} = P_{2j}$, then $C_j = P_{1j} = P_{2j}$;
2. if $P_{1j} \ne P_{2j}$, then
   - $C_j = P_{1j}$ with probability $p = \dfrac{f^s_{P_2}}{f^s_{P_1} + f^s_{P_2}}$
   - $C_j = P_{2j}$ with probability $1 - p$.

6.4   Mutation Operator and Variable Mutation Rate

After crossover, we apply the mutation operator by randomly altering a number of bit positions. The number of positions to mutate for a given solution depends on the mutation rate. We use the variable mutation rate of [1], which depends on the rate at which the GA converges. That is, lower mutation rates are used in early generations of the GA and mutation increases to higher rates when the population converges and then stabilizes to a constant rate afterwards. The mutation schedule below specifies the number B of bits to mutate [1].

$$B = \left\lceil \frac{m_f}{1 + \exp \frac{-4m_g(t - m_c)}{m_f}} \right\rceil \tag{14}$$

where t is the number of child solutions that have been generated so far, m_f specifies the final stable mutation rate, m_c is the number of solutions that should be generated such that the mutation rate is $\frac{m_f}{2}$, and m_g specifies the gradient at $t = m_c$. The value of m_f is user-defined and the values of m_c and m_g are problem-dependent parameters.

6.5   Heuristic Feasibility Operator

The solutions generated by crossover and mutation operators can be unfeasible. We propose a *heuristic feasibility operator* that maintains the feasibility of the solutions in the population and at the same time provides a local optimization method for fine-tuning the results of crossover and mutation operators. This operator consists of the last two phases of the DRC heuristic in Figure 1: *Construction* and *Reduction Phases* in this order. There is no need for *Initialization Phase* here since the GA has already generated a potentially good solution P_{sol}. The *Construction Phase*, however, starts with such a solution P_{sol} to generate a feasible solution. The feasibility operator is applied only for unfeasible solutions.

6.6   Initial Population

The initial population must consist of feasible solutions. Given the incidence matrix H and the initial probe set P, a random n-bit string $s_c = s_{c1} \ldots s_{cn}, 1 \le c \le N$, is generated with a high probability of being feasible, in the following way: we uniformly generate a random value $r \in [0,1]$ then set

$$s_{cj} = \begin{cases} 1 \text{ If } r \le D(p_j) \\ 0 \text{ Otherwise} \end{cases} 1 \le j \le n \qquad (15)$$

and, if necessary, we apply our heuristic feasibility operator to maintain the feasibility of s_c. This also has the advantage that the initial solutions are good solutions since the operator is an optimizer. Finally, given p_j, $D(p_j)$ can be considered as the probability of selecting p_j such that s_c is feasible.

6.7   Replacement Strategy

We use a *steady-state replacement strategy*. That is a newly generated feasible solution will replace a randomly chosen member of the population, which is usually a less fit individual. This strategy is also *elitist* since the best solutions are always kept and passed on to subsequent generations. Also, a GA with a steady-state replacement strategy converges faster than a GA with a *generational replacement strategy*.

6.8   The Algorithm

The GA can be summarized as follows:

1. Create a random population $\{S_1, \ldots, S_N\}$. Set $t \leftarrow 0$.
2. Select two solutions S_1, S_2 using fitness scaling and binary tournament.
3. Apply the fusion crossover on S_1, S_2 to yield C.
4. Mutate B randomly selected bits in C (B computed as in (14)).
5. Apply the heuristic feasibility operator on C.
6. If C is identical to any solution in the population, go to step 2; else, set $t \leftarrow t + 1$ and go to step 7.
7. Replace a randomly selected solution with an above-average fitness in the population by C (note that above-average here means *less* fit).
8. Repeat steps 2-7 until $t = M$ non-duplicate solutions have been generated (M is a user-defined parameter). The best solution in the population has the smallest fitness.

7   Computational Experiments

We performed experiments to show that evolutionary approaches are good alternatives to the solution of the non-unique probe selection problem and that they compare very well with current state-of-the-art heuristics. The programs were written in C and all tests ran on two Intel Xeon^{TM} CPUs 3.60GHz with 3GB of RAM under Ubuntu 6.06 i386.

7.1 Data Sets

We conducted experiments on ten artificial data sets (a1, ..., a5, b1, ..., b5 sets), and three real data sets (Meiobenthos, HIV-1 and HIV-2 sets). These data sets were used in all previous studies mentioned in Section 3 except for HIV-1 and HIV-2 sets which were used only in [5][6][9][10]. All these data sets along with their associated target-probe incidence matrices were kindly provided to us by Dr. Pardalos and Dr. Ragle [6]. Table 4 shows, in its second and third columns, the sizes $|T|$ and $|P|$ (number of targets and number of probes) of the incidence matrix for each set (M for Meiobenthos is the largest set). Due to space constraints, we refer the readers to [4][5][7] for the full details on the construction of these data sets.

7.2 Results and Discussions

All experiments were performed with parameters $c_{min} = 10$ and $s_{min} = 5$, as in all previous studies. The GA was run ten times on each data set; each time with a different random seed. Each run terminated when $M = 10,000$ non-duplicate solutions had been generated. The population size was $N = 100$, and the GA parameters were $m_f = 10, m_c = 200, m_g = 2$ (these values were obtained by trial-and-error).

Table 4 shows, for all data sets, the minimum sizes $|P_{min}|$ attained by the greedy heuristic of [7] (GrdS), the *Integer Linear Programming* (ILP) method of [3][4], the greedy heuristic of [5] (GrdM, this is the best greedy method described in literature for non-unique probe selection), our greedy heuristic previously introduced in [10] (DRC), the *Optimal Cutting-Plane* (OCP, this is the best method introduced in literature for non-unique probe selection) algorithm of [6], and our *Genetic Algorithm* (GA) discussed in [9]. Column V is the number of

Table 4. $|P_{min}|$'s of DRC, GA and other methods

Set	$|T|$	$|P|$	V	GrdS[7]	ILP[3][4]	GrdM[5]	DRC[10]	OCP[6]	GA[9] min	ave	max
a1	256	2786	6	1163	503	568	549	509	502	503.9 ± 1.3	506
a2	256	2821	2	1137	519	560	552	494	490	491.4 ± 0.7	492
a3	256	2871	16	1175	516	613	590	543	534	534.8 ± 1.0	537
a4	256	2954	2	1169	540	597	579	539	537	538.2 ± 0.6	539
a5	256	2968	4	1175	504	605	583	529	528	528.2 ± 0.4	529
b1	400	6292	0	1908	879	961	974	830	839	842.2 ± 2.0	845
b2	400	6283	1	1885	938	976	1013	842	852	854.8 ± 2.0	859
b3	400	6311	5	1895	891	951	953	827	835	838.7 ± 2.5	842
b4	400	6223	0	1888	915	1001	1019	873	879	882.5 ± 3.0	889
b5	400	6285	3	1876	946	1022	1019	874	890	892.8 ± 2.4	897
M	679	15139	75	3851	3158	2336	2084	1962	1962	1964.3 ± 2.5	1971
HIV-1	200	4806	20	-	-	531	487	451	450	450.7 ± 0.5	451
HIV-2	200	4686	35	-	-	578	506	479	476	477.7 ± 0.8	479

Table 5. Percentages of $|P_{\min}|$'s in relation to initial $|P|$'s

Data Sets & Sizes				Greedy Heuristics			Advanced Heuristics		
Set	$|T|$	$|P|$	V	GrdS[7]	GrdM[5]	DRC[10]	ILP[3][4]	OCP[6]	GA[9]
a1	256	2786	6	41.74	20.39	19.71	18.05	18.27	18.02
a2	256	2821	2	40.30	19.85	19.57	18.40	17.51	17.37
a3	256	2871	16	40.93	21.35	20.55	17.97	18.91	18.60
a4	256	2954	2	39.57	20.21	19.60	18.28	18.25	18.18
a5	256	2968	4	39.59	20.38	19.64	16.98	17.82	17.79
b1	400	6292	0	30.32	15.27	15.48	13.97	13.19	13.33
b2	400	6283	1	30.00	15.53	16.12	14.93	13.40	13.56
b3	400	6311	5	30.03	15.07	15.10	14.12	13.10	13.23
b4	400	6223	0	30.34	16.09	16.37	14.70	14.03	14.13
b5	400	6285	3	29.85	16.26	16.21	15.05	13.91	14.16
M	679	15139	75	25.44	15.43	13.77	20.86	12.96	12.96
HIV-1	200	4806	20	-	11.05	10.13	-	09.38	09.36
HIV-2	200	4686	35	-	12.33	10.80	-	10.22	10.16

additional virtual probes inserted to maintain the feasibility of the initial probe sets P. For the GA, we report for each data set, the minimum, average and maximum values over ten runs; for each run, these values are obtained from the last population of the GA and then averaged over ten runs.

In the table, each solution $P_{\min}$ contains the virtual probes inserted into it associated initial probe set P. Interestingly, we note that [6] reported $|P_{\min}|$'s minus the virtual probes but then compared with [3][4] whose $|P_{\min}|$'s include the virtual probes. In Table 5, we show the size of a solution $P_{\min}$ as a proportion of the initial candidate probe set P.

Our DRC heuristic greatly outperformed the greedy GrdS in all instances, and outperformed the greedy GrdM in 9 out of 13 instances. GrdM heuristic sorts the probes in decreasing order of the number of targets they hybridize, then selects probes in this order for satisfying the constraints, and finally, searches for redundant probes to remove. This probe sorting process is similar to selecting probes that cover the largest number of rows first, though this is not encoded in a selection function. GrdM uses no other information nor any selection function, thus it cannot differentiate between probes that hybridize to the same number of targets. Also, sorting does not guarantee that dominated targets are covered earlier in the probe selection process and, therefore, GrdM yields larger solutions than DRC, in general. DRC on the other hand encodes much more information about each probe in a selection function, Equation (13), and thus, is able to differentiate between good and bad probes. In particular, DRC performed particularly much better than ILP on the largest instance M. With regard to ILP, we note that [3][4] first applied the greedy GrdS heuristic to reduce an initial probe set P (and to reduce the ILP running time) and then further optimized the *reduced probe sets* with the ILP solver CPLEX (CPLEX is one of the leading mathematical programming software packages available and very few heuristic algorithms, if any, are able to compete with its results). CPLEX was *restricted* to search only a small portion of the solution space, hence ILP was not aware of the full initial probe sets. Our greedy DRC heuristic has no such restriction and runs faster.

The GA greatly outperformed the greedy heuristics in all instances. DRC improves GrdS by using a probe selection function (see Section 4) that guides the search for a minimal probe set. Note that DRC generates only a single solution at termination. Using the same selection function together with a population of solutions as well as genetic operators, The GA behaves as an implicit parallelization of DRC. The GA will be more appropriate than DRC for data sets much larger than the Meiobenthos data set. The GA performed substantially better than ILP in 9 out of 11 instances, and slightly outperformed OCP in 6 out of 13 instances. Unlike ILP, the GA and OCP are not restricted to search a small portion of the search space 2^P, and thus they can explore a much larger search space for near minimal solutions. In the worst cases, the GA still produces comparable results within at most +4.9% of ILP and +1.8% of OCP results (see Table 6). OCP uses integer linear programming principle and was proposed by [6] as an improvement (in time and minimization) of the ILP method of [3][4].

Table 6. Percentages of improvements of DRC and GA

Set	DRC[10]		GA[9]	
	GrdS[7]	GrdM[5]	ILP[3][4]	OCP[6]
a1	−52.8	−3.3	−0.2	−1.4
a2	−51.5	−1.4	−5.6	−0.8
a3	−49.8	−3.8	+3.5	−1.7
a4	−50.5	−3.0	−0.6	−0.4
a5	−50.4	−3.6	+4.8	−0.2
b1	−49.0	+1.4	+4.6	+1.1
b2	−46.3	+3.8	−9.2	+1.2
b3	−49.7	+0.2	−6.3	+1.0
b4	−46.0	+1.8	−3.9	+0.7
b5	−45.7	−0.3	−5.9	+1.8
M	−47.3	−10.8	−37.9	0.0
HIV-1	-	−8.3	-	−0.2
HIV-2	-	−12.5	-	−0.6

Table 6 reports the percentages of improvement, Imp%, of DRC over GrdS and GrdM, and of the GA over ILP and OCP. Imp% is computed as

$$\text{Imp} = \frac{P_{\min}^X - P_{\min}^Y}{P_{\min}^Y} \times 100 \ , \qquad (16)$$

where X is either DRC (compared with Y = GrdS or GrdM) or the GA (compared with Y = ILP or OCP). A negative (respectively, positive) value of Imp means that a X result is Imp% better (respectively, worse) than a Y result. Imp is negative when X returns a probe set smaller than $P_{\min}^Y$, therefore the smaller Imp is the better X is.

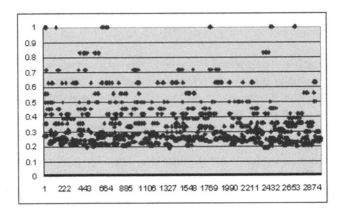

Fig. 2. $D(p)$ Distribution in a5 data set

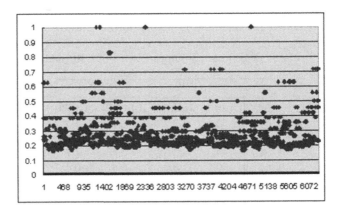

Fig. 3. $D(p)$ Distribution in b5 data set

Both DRC and the GA have difficulties in optimizing the sets b1–b5 than the sets a1–a5. We noticed that: 1) the a1–a5 sets contain a greater percentage of high-degree probes than the b1–b5 sets and 2) a greater percentage of probes in the b1–b5 sets have degrees $D(p) < 0.25$. In Figures 2 and 3, we show the distribution of the initial $D(p)$ values for a5 and b5 data sets. We found that 37% of the probes in the a5 set have degree $D(p) < 0.25$ and 65% of the probes in the b5 set have degree $D(p) < 0.25$. Similarly, there are more high-degree probes in a5 than in b5. In b5, not only too many probes have low degrees but also too many low-degrees probes are (almost) similar and DRC does not encode enough information to select between them; for such data sets, GrdM does better than DRC by selecting probes that cover the largest number of rows first. The GA behaves similarly to DRC, since it uses the selection function of DRC. Likewise, OCP as in GrdM focuses first on probes that cover the largest number of rows. Hence OCP does better than the GA on the b1–b5 sets. It remains to be seen if the GA will give better results on the b sets than it is giving currently, if given enough time.

As a final note, the GA is able to find good solutions at the expense of execution time. We did not compare the time performances of the methods reported here since we only implemented GA and DRC. The codes for the other methods were not available to us. The GA, however, is slower than all these methods; for example it ran for at least 2hrs (for the a data sets), 7hrs (for the b data sets) and 62 hrs (for the M data set). On the other hand, DRC ran for at most 80 seconds for all data sets.

8   Conclusions and Future Research

In this paper, we have discussed an evolutionary approach to the solution of the non-unique probe selection problem. We considered the case for single target separations only, not aggregated target separations. Experiments showed that evolutionary methods are able to obtain near minimal solutions at least comparable to the best known heuristics for this problem. One may argue that our evolutionary method is not very practical. However, since the probe set for microarray is only created once, the time spent to compute the minimal probe set is far less crucial than the size and quality of the probe set. We plan to study different variations of our heuristic feasibility operator since much speed-up should be expected by further improving this operator. Future research includes also the design of an evolutionary method that do not require a feasibility operator.

Acknowledgment. We thank Dr. Pardalos and Dr. Ragle for providing us with all the data sets used in this paper.

References

1. Beasley, J.E., Chu, P.C.: A genetic algorithm for the set covering problem. European J. Oper. Res. 94, 392–404 (1996)
2. Huang, Y., Chang, C., Chan, C., Yeh, T., Chang, Y., Chen, C., Kao, C.: Integrated minimum-set primers and unique probe design algorithms for differential detection on symptom-related pathogens. Bioinformatics 21, 4330–4337 (2005)
3. Klau, G.W., Rahmann, S., Schliep, A., Vingron, M., Reinert, K.: Integer linear programming approaches for non-unique probe selection. Discrete Applied Mathematics 155, 840–856 (2007)
4. Klau, G.W., Rahmann, S., Schliep, A., Vingron, M., Reinert, K.: Optimal robust non-unique probe selection using integer linear programming. Bioinformatics 20, i186–i193 (2004)
5. Meneses, C.N., Pardalos, P.M., Ragle, M.A.: A new approach to the non-unique probe selection problem. Annals of Biomedical Engineering 35(4), 651–658 (2007)
6. Ragle, M.A., Smith, J.C., Pardalos, P.M.: An optimal cutting-plane algorithm for solving the non-unique probe selection problem. Annals of Biomedical Engineering 35(11), 2023–2030 (2007)
7. Schliep, A., Torney, D.C., Rahmann, S.: Group testing with DNA chips: generating designs and decoding experiments. In: Proc. IEEE Computer Society Bioinformatics Conference (CSB 2003), pp. 84–91 (2003)

8. Tobler, J.B., Molla, M.N., Nuwaysir, E.F., Green, R.D., Shavlik, J.W.: Evaluating machine learning approaches for aiding probe selection for gene-expression arrays. Bioinformatics 18, s164–s171 (2002)
9. Wang, L., Ngom, A., Gras, R.: Non-Unique Oligonucleotide Microarray Probe Selection Method Based on Genetic Algorithms. In: Proc. 2008 IEEE Congress on Evolutionary Computation, Hong Kong, China, June 1-6 (to appear, 2008)
10. Wang, L., Ngom, A.: A model-based approach to the non-unique oligonucleotide probe selection problem. In: Second International Conference on Bio-Inspired Models of Network, Information, and Computing Systems (Bionetics 2007), Budapest, Hungary, December 10-13 (2007) ISBN: 978-963-9799-05-9

Stochastic π-Calculus Modelling
of Multisite Phosphorylation Based Signaling:
The PHO Pathway in Saccharomyces Cerevisiae

Nicola Segata and Enrico Blanzieri

Dipartimento di Ingegneria e Scienza dell'Informazione - University of Trento

Abstract. We propose a stochastic π-calculus modelling approach able to handle the complexity of post-translational signalling and to overcome some limitations of the ordinary differential equations based methods. The model we developed is customizable without *a priori* assumptions to every multisite phosphorylation regulation. We applied it to the multisite phosphorylation of the *Pho4* transcription factor that plays a crucial role in the phosphate starvation signalling in *Saccharomyces cerevisiae*, using available *in vitro* experiments for the model tuning and validation. The *in silico* simulation of the sub-path with the stochastic π-calculus allows quantitative analyses of the kinetic characteristics of the *Pho4* phosphorylation, the different phosphorylation dynamics for each site (possibly combined) and the variation of the kinase activity as the reaction goes to completion. One of the predictions indicates that the *Pho80-Pho85* kinase activity on the *Pho4* substrate is nearly distributive and not semi-processive as previously found analysing only the phosphoform concentrations *in vitro*. Thanks to the compositionality property of process algebras, we also developed the whole PHO pathway model that gives new suggestions and confirmations about its general behaviour. The potentialities of process calculi-based *in silico* simulations for biological systems are highlighted and discussed.

1 Introduction

Protein activity is regulated by a large set of post-translational modifications (PTMs) that often act cooperatively on different domains in order to combinatorially increase the possible states and behaviours of the target protein [1,2,3]. Among the possible PTMs, phosphorylation is very common and multisite phosphorylation is a frequent variant that enhances expressivity and enables non-linearity of protein responses [4,5,6]. The understanding of these kinds of regulations is crucial in many pathways, but experimental analysis of multisite phosphorylations remains complex even though some specific mass spectrometry based techniques were proposed [7,8]. For these reasons the experimental approach is very often coupled with computational models that are mandatory for an in-depth understanding of the kinetics of the regulation [9,10,11,12]. The modelling approach adopted in these cases is based on ordinary differential equations (ODE), which exhibit some limitations. First, some constrictive assumptions like

C. Priami et al. (Eds.): Trans. on Comput. Syst. Biol. X, LNBI 5410, pp. 163–196, 2008.
© Springer-Verlag Berlin Heidelberg 2008

the distributive and ordered behaviour of the multisite phosphorylation could be required to maintain the computational cost under a reasonable level [4]. Second, the system description is based on the concentrations which are continuous values while the biological interactions are discrete; this approximation is generally not acceptable when the number of interacting entities is small and when modifications not influencing the concentration occur. Third, the ODE are intrinsically deterministic and do not consider noise (unless artificially added like in [13]) and stochasticity at least in their natural formulation. Fourth, the ODE approach cannot handle all the possible analyses at the same time so it is necessary to focus on some partial aspects. So, up to now, neither the *in vitro* nor the ODE simulation approach can handle in the same experiment all the potentially interesting aspects in a quantitative way.

In this paper we present a modelling approach that overcomes the above limitations and introduces other desirable features. The modelling framework is based on process algebras theory used in biology for the first time in [14] and [15], reviewed in [16] and applied on small pathways and gene networks for example in [17,18,19,20]. We show that models built with these algebras can be tuned consistently with respect to general qualitative descriptions as well as quantitative values derived from biological experiments. Once a model is correctly validated, the *in silico* simulations can provide precise predictions of non-measurable biological aspects and new time-continuous analyses. Such models also permit the quantification of every possible single or multiple step transition of the systems. Regarding the multisite phosphorylation, it is possible, for example, to consider the quantification of binding events without phosphorylations that cannot be sensed by biological experiments because the phosphoform concentrations remain unchanged.

We modelled with the stochastic π-calculus (one of the most used process calculi, see Appendix A) the *Pho4* multisite phosphorylation which is the crucial event of PHO regulatory pathway in *Saccharomyces cerevisiae* [21]. The model was validated with respect to the three datasets of biological *in vitro* experiments taken from [9] that are the only quantitative information available for this kinase reaction. The simulations are performed with the SPiM simulator [22] briefly described in Appendix A, which is based on the Gillespie's algorithm [23].

We extended the validated *Pho4* multisite phosphorylation model to the whole phosphate condition signaling mechanism obtaining the PHO pathway model. This was possible relying on the compositionality (the ability of building models incrementally) of process algebras. The sub-paths and mechanisms added to the *Pho4* multisite phosphorylation model concern the dynamic regulation of processes numbers, the promoter activity, the transmembrane transport and protein synthesis and degradation. These modelled sub-paths are described as generally as possible and they represent a subset of a possible collection of basic stochastic π-calculus models that can be easily adapted to specific cases and combined together to obtain complex pathways or combination of pathways in a compositional way. The full PHO model can be only qualitatively validated because of the unavailability, at the best of our knowledge, of precise experimental data but permits to perform predictive analyses and observations about the overall behaviour of the pathway. We also verified the potentiality of the model to be predictive with respect to microarray data.

Our contribution on the phosphorylation PTM is a computational model able to handle every phosphorylation reaction with a variable number of phosphorylation sites. If the available biological values are precise enough to set properly the rates, our computational model can analyse and quantify complex and combined aspects that are usually not easily measurable or predictable.

The biological contributions of this work represent the first biological predictions obtained with process calculi approaches (and in particular with the stochastic π-calculus) and are relative to two aspects. a) On the *Pho4* multisite phosphorylation mechanism, we provide a model-based estimation of the average number of *Pho4* phosphorylations per *Pho80-Pho85* binding event with its variation during the reaction. This shows that the phosphorylation is more likely to be distributive than processive. Moreover, the *Pho4* site specific phosphorylation dynamics suggests quantitative estimation of the *Pho4* behaviour and further hypotheses on the kinetics of the target promoter regulations. b) On the whole PHO pathway we give some qualitative confirmations about the general mechanisms (like the importance of the *Pho81* feedback loop), we characterise the expression level of the species in different phosphate conditions and propose an analysis about the percentage of *Pho81* proteins active as inhibitors needed to signal the starvation and the intermediate phosphate conditions.

2   The PHO Pathway in Saccharomyces Cerevisiae

The PHO pathway is the regulatory mechanism of *Saccharomyces cerevisiae* responsible of the optimal use of inorganic phosphate when the availability of this nutrient in the medium is low, increasing the expression of multiple genes involved in its acquisition, uptake and storage (for a review refer to [21,24,25]).

The multi-domain phosphorylation of the *Pho4* transcription factor by the *Pho80-Pho85* cyclin-CDK kinase complex [26,27] is the key event of this signalling pathway. As shown for example in [28] and [29], the profiles produced by the phosphorylation of four of the five *Pho4* phosphorylation sites are associated with a precise and different behaviour of the transcription factor, favouring its extranuclear exportation, preventing its reimportation or inhibiting the cooperative binding with the *Pho2* transcription factor [30] on the promoters of *Pho5* [31] and *Pho81* [32,33] genes. The transnuclear membrane transportation of *Pho4* is carried out by the *Pse1* import receptor [34] and by *Msn5* export receptor [35] that recognise and binds only to a subset of its possible phosphorylation profiles. This partially redundant and overlapping regulation mechanism [36] results in multiple levels of possible responses to phosphate starvation and makes crucial the quantitative analysis of the 32 different phosphoform dynamics during the kinase reaction in order to understand precisely the *Pho4* behaviour in different conditions. The *Pho80-Pho85* kinase complex binds to the *Pho81* protein which becomes active as kinase inhibitor when the low concentration of phosphate

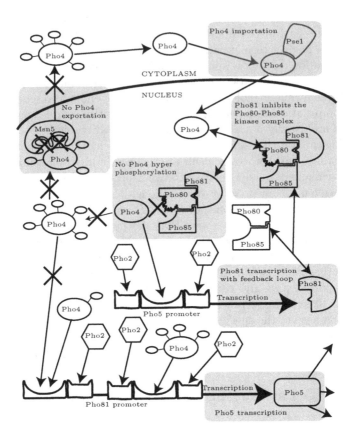

Fig. 1. A representation of the PHO pathway during phosphate starvation

causes the modification of its so called *minimum domain* (the way it happens is still unknown).

So phosphate starvation (Figure 1) results in the inhibition of the *Pho80-Pho85* complex by *Pho81* and this leads to the accumulation of *Pho4* in the nucleus and to the enhancing of the capability to bind to its target promoters cooperatively with *Pho2*. The list of the genes regulated directly by the PHO pathway (that includes *Pho5*, *Pho8* and *Pho84* and *Pho89*) comprehends the *Pho81* protein itself which thus performs a positive feedback loop in the pathway.

3   Stochastic π-Calculus Sub-models for the PHO Pathway

In this section we introduce the stochastic π-calculus models developed to simulate single sub-paths of the PHO pathway. First we describe a general multisite phosphorylation model, then we present the models for the dynamic regulation of the number of processes, for the promoter activity, for the transmembrane transport and for protein synthesis and degradation. The sub-models are described separately and are specified as general as possible because they have a general biological valence and can be useful for modelling other biological systems. The stochastic π-calculus is presented in Appendix A, in which are discussed the syntax and the semantics of the language as well as the SPiM simulator and some modelling abstractions that are used in this section. The π-calculus models are shown here with the non-stochastic formalism for clearness and with no assumptions about the distribution of the rates that will be associated to the actions, while the code fragments of the SPiM simulator language and the textual descriptions consider also the stochastic aspects.

3.1   The Multisite Phosphorylation Model

We describe here a model applicable to every multisite phosphorylation regulation of a general substrate with a desirable number of phosphorylation sites and without any *a priori* assumptions.

Figure 2 represents the possible evolutions of the system for a phosphorylation regulation with 3 different sites (s_1, s_2 and s_3). In the general, with n phosphorylation sites, there are 2^n states representing each possible substrate phosphorylation profile and 2^n states for each possible substrate bound to the kinase. Every kinase-substrate complex can disassociate in the two components or can perform a phosphorylation on a non already phosphorylated site. All the phosphorylation transitions are based on synchronous communications on private channels.

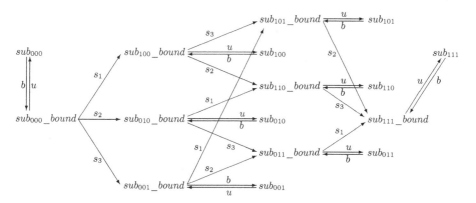

Fig. 2. All the possible phosphorylation and binding/unbinding transitions for a phosphorylation reaction with three phosphorylation sites. The free substrate is denoted with sub_{xxx}, the kinase-substrate complex with sub_{xxx}_bound and the phosphorylation profile xxx is defined by three binary digits (1 denotes phosphorylation). The arrows represent transitions and their labels indicate the associated event (with s_i the phosphorylation event on site s_i, b the kinase-substrate binding event and u the kinase-substrate unbinding event).

---

1 ν *sub_kin_bind*

2 (! sub_{100}().ν *unbind* ν s_2 ν s_3

3   (*sub_kin_bind*(*kin_unbind*).

4   (*unbind*⟨⟩.*kin_unbind*⟨⟩.**0**

5 | s_2⟨⟩.**0**

6 | s_3⟨⟩.**0**

7 | sub_{100}_bound⟨*unbind*, s_2, s_3⟩.**0**))

8 | ! sub_{100}_bound(*unbind*, s_2, s_3).

9   (*unbind*().sub_{100}⟨⟩.**0**

10   + s_2().sub_{110}_bound⟨*unbind*, s_3⟩.**0**

11   + s_3().sub_{101}_bound⟨*unbind*, s_2⟩.**0**)

12 | ! *kin*().ν *kin_unbind*

13 (*sub_kin_bind*⟨*kin_unbind*⟩.*kin_bound*⟨*kin_unbind*⟩.**0**)

14 | ! *kin_bound*(*kin_unbind*).*kin_unbind*().*kin*⟨⟩.**0**)

Fig. 3. A simplified version of the multisite phosphorylation model

In the description of the stochastic π-calculus model we consider a substrate with three phosphorylation sites and we focus only on one of the 8 possible phosphorylation profiles, namely the one with the first site already phosphorylated (sub_{100} and sub_{100}_bound for the bound version). The other profiles have the same structure and phosphorylation mechanism.

A first approximation of the model is shown in Figure 3. It contains processes of the free substrate (only the free substrate with the 100 phosphorylation pattern is shown, lines 2-7), of the substrate bound to the kinase (only the bound substrate with the 100 phosphorylation pattern is shown, lines 8-11) and of the kinase (lines 12-14). Notice that we adopt the state identifiers shown in Figure 2 for the channels guarding the corresponding processes since there is no possibility of misinterpretation. The binding between the kinase and the substrate occurs through the *sub_kin_bind* (a global channel declared in line 1) on which the channel for the unbinding (*kin_unbind* line 12) is transmitted. The binding rate is associated to the *sub_kin_bind* channel while the unbinding occurs instantly (i.e. with an infinite rate) after a stochastic delay internal to the bound substrate process. This delay is implemented with a communication on a channel restricted to the substrate (*unbind* declared in line 2) because it can be passed through the substrate sub-processes after the effective binding (line 7) or a phosphorylation (line 10 or 11), while the delay modelled with a silent action would be natively restricted to a single process. The unbinding signal is handled by the substrate and not by the kinase because the binding duration depends also on the phosphorylation profile of the substrate which is not known by the kinase. In addition to the *unbind* channel the parameters passed to the bound substrate process are the channels for the phosphorylation on the sites that are not already phosphorylated (in this case s_2 and s_3). These three passed channels, on which three sub-processes of the substrate in parallel are willing to perform an output (lines $4-6$), are used by the correspondent bound version of the kinase (lines $8-11$) to perform the unbinding or the phosphorylation on s_2 or s_3. More precisely when a synchronization occurs on the *unbind* channel the execution returns to the free version of the kinase (sub_{100}), when a communication occurs on the s_2 channel the bound kinase process with s_2 phosphorylated is activated (sub_{110}_bound not shown here) and similarly for the communication on the s_3 channel.

Two properties of the model are crucial for expressing the correct biological behaviour of the multisite phosphorylation:

- The interactions on the phosphorylation and unbinding channels are in competition and only one can occur in a single kinase-substrate binding event. The concurrent activity of the channels is assured by the parallel composition of the processes willing to perform a communication on the phosphorylation and unbinding channels (lines 4-6), while the competitive behaviour and the uniqueness of the performed communication is guaranteed by the semantics of the non-deterministic choices that compose the receiving processes (lines 9-11).
- The phosphorylation of a site is independent from the others. This is assured because, when a phosphorylation occurs (lines 10 or 11), the correspondent new bound version of the substrate is activated passing as parameters the other phosphorylation channels on which the sending operation is still willing to be performed since the kinase-substrate binding event.

The model in Figure 3 is complete for a qualitative description but presents some problems from an effective simulation point of view because some sub-processes remain active willing to perform an operation on some private channel that will never have a correspondent complementary communication operation. In a quantitative simulation framework this results in a huge and useless computational and memory effort. In particular, when one of the branches of the non-deterministic choice of the bound substrate (lines 9-11) is selected denoting a phosphorylation, the other branches are removed and so the parallel sub-processes of the substrate that are willing to perform an output on the other phosphorylation channels (lines 5-6) are blocked forever.

The problem is tackled organising each sending operation on a phosphorylation channel in a non-deterministic choice structure with a new sub-process and forcing, if the phosphorylation does not occur, the execution of the other branch when the substrate and kinase disassociate.

The refined model is shown in Figure 4; the *end_s_2* and *end_s_3* channels (line 2) are added to send the termination signal to the sub-processes blocked for the phosphorylation signal (lines 5 and 6) after the unbinding event which occurs on the *unbind* channel (line 4). Note that these two sub-processes have the receiving operation on the termination channels not only on the right branch of the non-deterministic choice, but also on the left one because if a phosphorylation occurs, the termination signal could no more be received and so the process would be blocked on the sending of the termination signal (line 4).

The code fragments of the SPiM implementation of the multisite phosphorylation model is shown in Figure 5.

$\nu \, sub\_kin\_bind$
$(\, !\, sub_{100}().\, \nu \, unbind \; \nu \, s_2 \; \nu \, s_3 \; \nu \, end\_s_2 \; \nu \, end\_s_3$
$(\, sub\_kin\_bind(kin\_unbind).$
$\qquad (\, unbind\langle\rangle.end\_s_2\langle\rangle.end\_s_3\langle\rangle.kin\_unbind\langle\rangle.\mathbf{0}$
$\qquad |\; s_2\langle\rangle.end\_s_2().\mathbf{0} + end\_s_2().\mathbf{0}$
$\qquad |\; s_3\langle\rangle.end\_s_3().\mathbf{0} + end\_s_3().\mathbf{0}$
$\qquad |\; sub_{100}\_bound\langle unbind, s_2, s_3\rangle.\mathbf{0}\,)\,)$
$|\, !\, sub_{100}\_bound(unbind, s_2, s_3).$
$\qquad (\, unbind().sub_{100}\langle\rangle.\mathbf{0}$
$\qquad + s_2().sub_{110}\_bound\langle unbind, s_3\rangle.\mathbf{0}$
$\qquad + s_3().sub_{101}\_bound\langle unbind, s_2\rangle.\mathbf{0}\,)$
$|\, !\, kin().\nu \, kin\_unbind$
$(\, sub\_kin\_bind\langle kin\_unbind\rangle.kin\_bound\langle kin\_unbind\rangle.\mathbf{0}\,)$
$|\, !\, kin\_bound(kin\_unbind).kin\_unbind().kin\langle\rangle.\mathbf{0}\,)$

Fig. 4. The multisite phosphorylation model

Model Parameters. The rates associated to the channels of the model that are biologically meaningful are those regarding the site-specific phosphorylation rates and the kinase-substrate association and disassociation rates. These are the rates that need to be estimated and tuned consistently with the biological data. Apart for the phosphorylation rates, the model permits to set different values for each phosphorylation profile. All the other rates have only technical modelling purposes without biological meanings and all have infinite rate.

```
...
new sub_kin_bind@sub_kin_bind_rate : chan(chan)
...
and sub100() = (
    new unbind@unbinding_rate_100 : chan()
    new s2@s2_rate : chan()
    new s3@s3_rate : chan()
    new end_s2 : chan()
    new end_s3 : chan()
    ?sub_kin_bind(kin_unbind);
        ( !unbind;!end_s2;!end_s3;!kin_unbind;()
        | do        !s2;?end_s2;()    or   ?end_s2;()
        | do        !s3;?end_s3;()    or   ?end_s3;()
        | sub100_bound(unbind,sp2,sp3)   )    )
...
and sub100_bound(unbind:chan,s2:chan,sp3:chan) = (
    do        ?unbind;sub100()
        or   ?s2;substrate_bound_110(unbind,s2)
        or   ?s3;substrate_bound_101(unbind,s1)     )
...
and kin() = (
    new unbind : chan()
    sub_kin_bind(unbind);?unbind;kin()   )
...
```

Fig. 5. The SPiM fragments of the model in Figure 4

3.2   Dynamic Regulation of the Number of Processes

The stochastic π-calculus does not permit the direct modelling of the environmental conditions. Since in biology some modifications are concentration-driven, this characteristic can be limiting. We propose to model the partial modification of a set of protein caused by precise concentration conditions imposing artificially the percentage of modified proteins. If a relation between the percentage of modified proteins and the concentration conditions can be established, the limitation of the calculus becomes less crucial. We must however consider that the number of protein of a species can change in time and so the number of modified proteins must change dynamically.

For these reasons, we propose here a general model for regulating the relative amount of a certain set of processes in a state S_1 with respect to another set of processes in state S_2 and with possibility of adding or removing new processes in a state and of changing the state. The basic mechanism is represented in Figure 6 in which it is implicit that a process waiting on a channel A is considered to be in state A; a process in S_1 can change the state with a rate r_2 associated with the silent action τ_2 and a process in S_2 can do the opposite with a rate r_1 associated with τ_1.

$$\boxed{\quad 1 \quad !\,S_1().\,(\tau_2.\,S_2\langle\rangle) \ \mid \ !\,S_2().\,(\tau_1.\,S_1\langle\rangle) \quad}$$

Fig. 6. The basic model for the process number regulation

The ratio between the two rates that regulate the two branches of the parallel composition is responsible for the balance of the number of the two processes. Intuitively, if we want the number of processes in S_1 to be n times greater than those in S_2 it is necessary that $r_1/r_2 = n$.

Denoting with P_{S_1} the percentage of processes in S_1 with respect to the total number of processes in S_1 and in S_2 (and similarly for P_{S_2}) it is required that:

$$\begin{cases} r_1 \times P_{S_2} = r_2 \times P_{S_1} \\ P_{S_1} + P_{S_2} = 100 \end{cases}$$

Defining $r = r_1 + r_2$ we can derive the value of r_1 and r_2 with respect to r in the following way:

$$r_1 = \frac{P_{S_1}}{100} \times r \quad r_2 = \frac{P_{S_2}}{100} \times r \qquad with \quad P_{S_1} + P_{S_2} = 100 \tag{1}$$

With this assumption and with a sufficient number of processes in S_1 and in S_2 and possibly after a convergence time, the framework assures that the percentage of processes in S_1 and in S_2 with respect to the total number of processes in the framework converges respectively to P_{S_1} and P_{S_2}. This fact is formally demonstrated converting the model in a continuous time Markov chain and finding the limiting probabilities (see Appendix B).

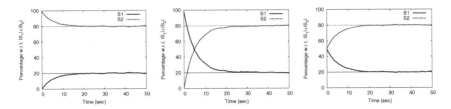

Fig. 7. These three plots represents the modifications during the simulations of the number of processes in the two states (S_1 and S_2). In this case $P_{S_1} = 80\%$ (and so $P_{S_1} = 20\%$), $r = 1$ and consequently $r_1 = 0.8$ and $r_2 = 0.2$. We can observe that the percentages of the number of processes in the two states approach P_{S_1} and P_{S_2} and that the percentages at the steady state are independent from the initial percentages.

Some properties of the model:

- The percentage of process in S_1 or in S_2 after a convergence time is independent from the initial concentration of the two processes as shown experimentally in Figure 7.
- The noise at equilibrium heavily depends on the number of processes.
- The convergence time depends on $r = r_1 + r_2$.
- Higher the r, higher the accuracy of the framework, but also the number of transitions and thus the computational effort. Note that the value of r must be chosen before forcing the values of r_1 and r_2 with the equations (1).

However, this model has some limitations like the fact that the regulator and regulated processes are the same and that only a single process definition can belong to a state.

In Figure 8 we propose an extension in which the two states are *active* (A) and *inactive* (I). The regulated processes must have the following structure:

$$!P\_act().(\,\tau^1.P\_act\langle\rangle + \tau^2.P\_act\langle\rangle + \ldots + \tau^n.P\_act\langle\rangle\,)$$

with $n \geq 1$ and where each τ^j, with $j \in \{1\ldots n\}$, represents every possible process with a non infinite and reasonably short execution time. To be regulated this process is decomposed putting in parallel each non-deterministic choice branch and embedding it in a communication based on the received *end* channel as follows:

$$
\begin{array}{ll}
1 & !\,P^1\_act(end).\tau^1.end\langle\rangle.\mathbf{0} \\
2 & |\,!\,P^2\_act(end).\tau^2.end\langle\rangle.\mathbf{0} \\
3 & |\,\ldots \\
4 & |\,!\,P^n\_act(end).\tau^n.end\langle\rangle.\mathbf{0}
\end{array}
$$

These branches are called from the $P_i\_act()$ process of line 4 of the model; note the index i which denotes the possibility to have more than one type of process under regulation. The original model shown Figure 6 acts here (lines 16-19) as the engine of the regulation but it is separated by the regulated processes; for every regulated process exists a A or I process reflecting its state. The engine works sending an empty communication on the global *activate* channel to activate an inactive regulated process on which it also synchronizes the state changing from inactive ($I()$) to active ($A()$). When the state moves from active to inactive a sending operation with an infinite rate is performed on the *inactivate* channel that must be received from the corresponding non-deterministic choice branch (line 7) of one of the real processes in the active state ($P_i\_act()$).

The engine processes ($A()$ and $I()$) contain also the receiving operation for the termination of an active (or inactive) process in the non-deterministic choice structure; the signal is sent by a specific process ($P_i^{end}\_act()$ or $P_i^{end}\_inact()$, lines 22 and 23) and, after forwarding the termination signal to the regulated processes, the process $A()$ or $I()$ dies. To add a new process in the regulation framework it is sufficient to call the $P_i^{init}\_act()$ or $P_i^{init}\_inact()$ (lines 20 and 21) process according to the initial state we want to impose to the process. The $P_i^{init}\_act()$ process simply starts a new active process of the engine ($A()$) in parallel with the regulated process in its active state (and similarly $P_i^{init}\_inact()$).

The model, shown in Figure 8, maintains the same properties of the one shown in Figure 6 but can be applied to an arbitrary number of different process types.

$$
\begin{aligned}
&1\quad \nu\, P_i\_act \;\; \nu\, P_i\_inact \;\; \nu\, A \;\; \nu\, I \;\; \nu\, P_i^{init} \;\; \nu\, P_i^{init} \\
&2\quad \nu\, P_i^{end}\_act \;\; \nu\, P_i^{end}\_inact \;\; \nu\, activate \;\; \nu\, inactivate \\
&3\quad \nu\, active\_end \;\; \nu\, inactive\_end \\
&4\quad (\,!P_i\_act().(\;\; \nu\, end \;\; P_i^1\_act\langle end\rangle.end().\,P_i\_act\langle\rangle \\
&5\qquad\qquad + \ldots \\
&6\qquad\qquad + \nu\, end \;\; P_i^n\_act\langle end\rangle.end().\,P_i\_act\langle\rangle \\
&7\qquad\qquad + inactivate().\,P_i\_inact\langle\rangle \\
&8\qquad\qquad + active\_term().\mathbf{0}\,) \\
&9\quad |\,!P_i\_inact().(\;\; \nu\, end \;\; P_i^1\_inact\langle end\rangle.end(). \\
&10\qquad\qquad\qquad P_i\_inact\langle\rangle \\
&11\qquad\qquad + \ldots \\
&12\qquad\qquad + \nu\, end \;\; P_i^n\_inact\langle end\rangle.end(). \\
&13\qquad\qquad\qquad P_i\_inact\langle\rangle \\
&14\qquad\qquad + activate().\,P_i\_act\langle\rangle \\
&15\qquad\qquad + inactive\_term().\mathbf{0}\,) \\
&16\quad |\,!A().(\;\; \tau_2.\,inactivate\langle\rangle.\,I\langle\rangle \\
&17\qquad\qquad + active\_end().\,active\_term\langle\rangle\,) \\
&18\quad |\,!I().(\;\; \tau_1.\,activate\langle\rangle.\,A\langle\rangle \\
&19\qquad\qquad + inactive\_end().\,inactive\_term\langle\rangle\,) \\
&20\quad |\,!P_i^{init}\_act().(A\langle\rangle\,|\,P_i\_act\langle\rangle) \\
&21\quad |\,!P_i^{init}\_inact().(I\langle\rangle\,|\,P_i\_inact\langle\rangle) \\
&22\quad |\,!P_i^{end}\_act().\,active\_end\langle\rangle \\
&23\quad |\,!P_i^{end}\_inact().\,inactive\_end\langle\rangle\,)
\end{aligned}
$$

Fig. 8. The model for the dynamic regulation of the number of processes in two classes (active and inactive)

3.3   The Promoter Model

The gene transcription regulation is one of the powerful mechanisms through which a pathway can modulate its responses. We propose a simple model for describing the promoter activity that is regulated by the transcription factors which bind to specific regions of the promoter. When all the needed transcription factors are bound to the promoter, it allows the gene transcription and the unbinding of the bound transcription factors.

The model shown in Figure 9 represents a promoter that needs the binding with three different transcription factors to transcribe the corresponding gene. We model here the biological mechanisms through which the transcribed $mRNA$ results in the target protein with a single stochastic delay.

Theoretically, the number of different processes needed to represent each possible binding profile for modelling a promoter regulated by n transcription factors is 2^n. However the representation of all the states is redundant since the only important state is the one with all the transcription factors bound because it represents the only condition that allows the gene transcription. For this reason the model is made up of the parallel composition of n sub-processes for the binding of the n transcription factors (lines 4-6) and of the process representing the transcription and the reconfiguration of the promoter after the transcription (lines 7-10). The trick used to avoid the explosion of explicit process states consists in guarding the transcription process with n inputs in sequence that are fired one by one by the transcription factor bindings. More precisely the guard consists in n private channels called ok (declared in line 3 and used as guards on line 7) that are waiting for an input before the transcription process; a step of the guard is fired when one of the transcription factors has performed the binding through the specific public channel (tf_1, tf_2 or tf_3) and the corresponding branch of the parallel composition has notified the event with an output on the ok channel. When the guard is completely consumed the transcription is enabled and the model starts the process of the new protein (line 8) after a transcription delay. In parallel with the new protein activation the promoter must perform the unbinding of the transcription factors, the reconfiguration delay and the recursive call to the promoter process. The unbindings

```
1    ν tf₁  ν tf₂  ν tf₃
2    !promoter().
3        ( ν ok  ν tf₁ᵒᵘᵗ  ν tf₂ᵒᵘᵗ  ν tf₃ᵒᵘᵗ
4            ( tf₁⟨tf₁ᵒᵘᵗ⟩. ok⟨⟩. 0
5            | tf₂⟨tf₂ᵒᵘᵗ⟩. ok⟨⟩. 0
6            | tf₃⟨tf₃ᵒᵘᵗ⟩. ok⟨⟩. 0
7            | ok(). ok(). ok(). τ_transcription.
8                ( τ_prot_cod. new_prot⟨⟩
9                | tf₁ᵒᵘᵗ⟨⟩. tf₂ᵒᵘᵗ⟨⟩. tf₃ᵒᵘᵗ⟨⟩.
10                  τ_reconfig. promoter⟨⟩ ) ) )
```

Fig. 9. The promoter model

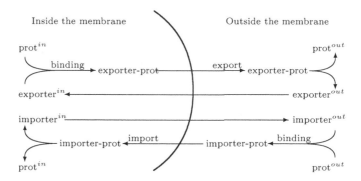

Fig. 10. The schematic behaviour of the transmembrane transporting mechanisms

occur through specific private channels (tf_1^{out}, tf_2^{out} and tf_2^{out} declared in line 3) passed to the transcription factors at the binding time and are tuned with an infinite rate.

3.4   The Transmembrane Transport Model

A simple general structure for modelling the protein exportation and importation through a membrane is described here. The model can be applied to membrane receptors or to general transport proteins. The schematic behaviour of the importer and of the exporter are shown in Figure 10; notice that, with the proposed level of abstraction, the behaviour of a transport carried out by a membrane receptor or by a non membrane transporter can be distinguished only by the binding and transporting rates, since in the π-calculus no spatial information is taken into account. The π-calculus sub model for the exporter is shown in Figure 11; the model for the importer is similar. The exporter (lines 2-6) binds to the protein performing an output on the *prot_binding* channel (line 1) on which it sends also the *binding_end* channel (line 3) associated with an infinite rate for signalling the unbinding after the transport. A particular branch of the protein process (line 12) receives the binding signal and immediately performs an output on the unbinding private channel which can be received by the exporter only after performing a silent action simulating the transport delay. During the silent action both the exporter and the protein cannot perform other actions thus simulating a real binding even if a specific process is not instantiated. After the unbinding the protein calls the process denoting its extra-membrane location characteristics ($prot^{out}$ not shown here) while the exporter, after a delay for the reimportation or reconfiguration process, performs a recursive call to itself.

3.5   The Protein Synthesis and Degradation Model

Very often the number of the species in the model is considered constant in the computational specification of biological systems. For this reason the modelled proteins have no synthesis and degradation mechanisms. However, when the

$$
\begin{array}{ll}
1 & \nu\, prot\_binding \\
2 & (\,!\, exporter(). \\
3 & \quad (\,\nu\, binding\_end \\
4 & \quad\quad (\, prot\_binding\langle binding\_end\rangle. \\
5 & \quad\quad \tau_{transport}.\, binding\_end(). \\
6 & \quad\quad \tau_{import}.\, exporter\langle\rangle\,)\,) \\
7 & |\,!\, prot\langle\rangle. \\
8 & \quad (\, sub\_process\_1\langle\rangle \\
9 & \quad +\, sub\_process\_2\langle\rangle \\
10 & \quad +\, \ldots \\
11 & \quad +\, sub\_process\_n\langle\rangle \\
12 & \quad +\, prot\_binding(unbinding). \\
13 & \quad\quad unbinding\langle\rangle.\, prot^{out}\langle\rangle\,)\,)
\end{array}
$$

Fig. 11. The transmembrane transport model

$$
\begin{array}{ll}
1 & !\, prot\_synth().\tau_{synth\_rate}. \\
2 & \quad \nu\, degradation \\
3 & \quad (\, prot\_synth\langle\rangle \\
4 & \quad |\, degradation\langle\rangle \\
5 & \quad |\, prot\langle degradation\rangle\,) \\
6 & |\,!\, prot(degradation). \\
7 & \quad (\,\ldots \\
8 & \quad +\, degradation().\, \mathbf{0}\,)
\end{array}
$$

Fig. 12. The protein synthesis and degradation model

studied pathway presents some non constitutively expressed protein, we must consider synthesization. As a consequence, it is necessary to have a degradation mechanism, otherwise the number of the synthesized proteins grows systematically. However, also in the case of constitutively expressed proteins the synthesis/degradation mechanism can have a crucial role, mainly due to the biological noise that affects the protein number in equilibrium conditions and to the fact that the synthesized proteins do not present PTM modifications. The model is shown in Figure 12; the synthesis process (lines 1-5) consists in a delay reflecting the synthesis rate after which the process of the new protein is launched in parallel with a recursive call of the synthesis process. The degradation is based on a rate (associated to the *degradation* channel declared on line 2) reflecting the half-life value of the protein; the synthesized protein process is called (line 5) passing as argument the degradation channel on which a parallel sub-process is willing to perform an output (line 4). The new protein process has a sub-process (line 8) in non-deterministic choice with all the other sub-processes that is willing to receive an input on the degradation channel after which the *prot* process terminates.

Note that the synthesis process can be embedded in other processes like the case of the promoter modelling. The new protein process in Figure 12 is not defined very precisely; in general, it must be assured that if another branch is selected a sub-branch for receiving the degradation signal is present and that if another process is called, the degradation channel is passed and handled similarly.

4   The *Pho4* Multisite Phosphorylation Model

The stochastic π-calculus multisite phosphorylation approach previously described is easily customizable to model the multi-domain phosphorylation of the *Pho4* transcription factor by the *Pho80-Pho85* cyclin-CDK kinase complex. Apart from the number of phosphorylation sites (five) the quantitative parameters of the model that need to be tuned are those regarding the different phosphorylation rates and the binding time between *Pho4* and *Pho80-Pho85* kinase complex with respect to the *Pho4* phosphorylation configuration. The model can set a different unbinding rate for every possible phosphorylation profile of *Pho4* but we choose to use only the five rates representing the mean of each possible state of the five different phosphoforms to be consistent with the existing biological data; the binding and unbinding rates are retrieved directly from [9] that reproduces the *Pho4* phosphorylation reaction *in vitro*. The same work quantifies three biological aspects: a) the phosphoform concentration variations, b) the site preferences of the substrate calculated as the amount of each phosphopeptide as the percentage of the total phosphorylation for every phosphoform and c) the average number of phosphorylations per binding event without considering the binding with no phosphorylations. These data give enough information to infer the phosphorylation rates; we artificially set the rates and observe how they fit the experimental data. If the fitness is not satisfying, we modify some rates iterating the process until a reasonably good configuration is found. Notice that, in general, it is not assured neither that the process eventually converges nor that the available data give enough constraint to the system to apply the process. In the particular case of setting the phosphorylation rates of *Pho4* the process is relatively simple because of the multidimensionality of the available data: the site preferences define the ratios between the phosphorylation rate of each site while the site preferences and the average phosphorylation influence the absolute "scale" of the values.

5   The PHO Pathway MODEL

The whole modelling of the PHO pathway is developed applying the described stochastic π-calculus sub-models to single PHO sub-paths; this approach is made effective by the compositionality property of process algebras. In particular the sub-models are applied as follows:

– The multisite phosphorylation model (Section 3.1) is applied to the reaction between *Pho4* and *Pho80-Pho85* as discussed in the previous section. The general sub-model is also extended with a simple dephosphorylation mechanism that is as general as possible because of the lack of biological information about this aspect. The dephosphorylation is implemented inside the *Pho4* processes and applied to the not bound substrate profiles both inside and outside the nucleus. It consists in some branches of the non-deterministic choice (one for each phosphorylated site) that after a delay regulated by a rate perform the dephosphorylation of a site calling the *Pho4* form with the specific site not phosphorylated. Similarly to the phosphorylation mechanism, the dephosphorylation rates can be set with site preferences.

– The framework to dynamically regulate the number of processes (Section 3.2) is applied to the *Pho81* cyclin-dependent kinase inhibitor to maintain a constant percentage of active inhibitors also with a variable total number of *Pho81* proteins. This because the phosphate concentration cannot directly be integrated in the model and the effect of the concentration is the level of *Pho81* active as *Pho80-Pho85* inhibitors.

– The promoter modelling (Section 3.3) is applied to the promoter of *Pho5* and *Pho81*. The first consists in five binding sites subdivided in 2 regulatory elements (*UASp1* and *UASp2*); three sites are for the *Pho2* binding and two for the *Pho4* binding. The promoter of *Pho81* has only one regulatory element (*UASp2*). Notice that the *Pho2* transcription factor can always bind to the promoters, while *Pho4* must be nuclear and with particular phosphorylation profiles.

– The transmembrane transport modelling (Section 3.4) is applied to the extranuclear exportation of *Pho4* with *Msn5* and to the nuclear reimportation with *Pse1*. Both the importer and the exporter recognize and bind only to a subset of the phosphorylation profiles.

– The synthesis and degradation mechanisms (Section 3.5) are applied to *Pho4* (for its influence in the phosphorylation dynamics) and to the non constitutively expressed protein *Pho5* and *Pho81*

Some of the sub-models are applied to the same species or mechanism; *Pho4* is involved in the transportation, phosphorylation, synthesis and degradation while the promoters integrate the synthesis process. In these cases the compositional approach is not immediate from a modelling point of view; however, if the processes are based on the non-deterministic choice, very often the only needed operation is the union of the two structure representing different behaviours of the same entity. Figure 13 shows a fragment of the PHO pathway model representing the *Pho4* transcription factor in one of the 32 phosphorylation profiles. The structure is the non-deterministic composition of the phosphorylation (first branch of the more external non-deterministic choice), degradation (second branch), dephosphorylation (third, fourth and fifth branches) and extranuclear exportation (the last three branches) mechanisms. Moreover, the degradation operation needs that the degradation channel is passed to every activated subprocess. The protein abundances are taken from [37] and [38] and scaled by a factor of 10 for computational reasons; the model is intrinsically not linear, but

```
and pho4_11010(degradate:chan) = (
 new unbind@bound_time_phospho3_11010 : chan()
 new degradation : chan()
 new s3@s3_c : chan()
 new s6@s6_c : chan()
 new end_s3 : chan()
 new end_s6 : chan()
 (do    ?pho80pho85pho81_pho4_bind
            (pho80pho85pho81_pho4_bind_out);bind();
     (
        do    !unbind;!pho80pho85pho81_pho4_bind_out;
              !end_s3;!end_s6;()
            or ?degradation;!pho80pho85pho81_pho4_bind_out;
              !end_s3;!end_s6;()
    |  do    !s3;?end_s3;() or ?end_s3;()
    |  do    !s6;?end_s6;() or ?end_s6;()
    |  pho4_11010_bound
              (degradate,degradation,unbind,s3,s6)
     )
 or ?degradate;()
 or delay@dephospho_s1_rate;pho4_01010(degradate)
 or delay@dephospho_s2_rate;pho4_10010(degradate)
 or delay@dephospho_s4_rate;pho4_11000(degradate)
 or ?UASp1s1(unbindP:chan);?unbindP;pho4_11010(degradate)
 or ?UASp2s2(unbindP:chan);?unbindP;pho4_11010(degradate)
 or ?UASp1s1b(unbindP:chan);?unbindP;pho4_11010(degradate)))
```

Fig. 13. The SPiM fragments of the stochastic model of the *Pho4* transcription factor when it is nuclear and phosphorylated on sites s_1, s_2 and s_4

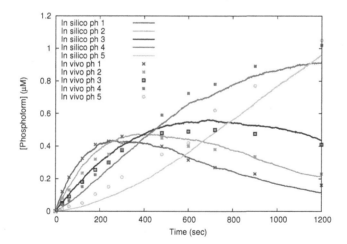

Fig. 14. The *Pho4* multisite phosphorylation model validation. The comparison between *in vitro* and *in silico* on the phosphoform concentrations among the reaction course; the points represent the *in vitro* experimental values taken from [9] (Figure 5a), the curves the continuous-time *in silico* simulations.

some simulations with different scaling factor suggest that the overall qualitative behaviour of the system is reasonably invariant for factors lower than 20.

6   In Silico Simulations vs. Biological Data

The model simulations were performed with SPiM [22] and basic Montecarlo statistical methods were applied to the analyses to improve robustness and precision of the computational results. The biological outcomes of the *in silico* analyses are listed in the following.

6.1   *Pho4* Multisite Phosphorylation Model Tuning and Validation

The model fit well the three *in vitro* experimental datasets taken from [9]. The phosphoforms time dependent quantification is reported in Figure 14 while Figure 15 points out the differences with respect to the site preferences of site s_2, s_3, s_4 and s_6 (s_1 is not taken in account because no *in vitro* data are available) between the traditional and the new proposed approach. In [9] the average number of phosphorylations for binding event is calculated simply multiplying the percentage of each phosphoform by the number of times it was phosphorylated (we call this calculation *ExpAvgPho*). The model based estimation of *ExpAvgPho* is very similar to biological data: the *in vitro* experiments of [9] give 2.06 while the output of the *in silico* experiments is 2.05.

6.2   *In silico* Estimation of the Average Number of *Pho4* Phosphorylation Events Per Binding Event between *Pho4* and the Cyclin-CDK *Pho80-Pho85*

The average number of phosphorylations for binding event computed following the *ExpAvgPho* calculation does not consider the bindings with no phosphorylations, leading to an overestimation of the phosphorylation activity. The *Pho4* multisite phosphorylation model is developed respecting the independence between the kinase-substrate binding and each site-specific phosphorylation and removing the phosphorylation channels of the sites on which an interaction occurred. In this way the model natively prevents attempts to phosphorylate sites

	s_2 site			s_3 site			s_4 site			s_6 site		
	bio	sπ	err	bio	sπ	err	bio	sπ	err	bio	sπ	err
Phosphoform 1	12.0	12.2	0.2	12.0	12.1	0.1	16.1	15.2	0.7	59.9	60.5	0.6
Phosphoform 2	17.1	16.9	0.2	17.1	16.7	0.4	22.0	20.9	1.1	43.8	45.5	1.7
Phosphoform 3	20.6	20.4	0.2	20.6	20.5	0.1	24.4	23.9	0.5	34.4	35.2	0.8
Phosphoform 4	23.0	23.1	0.1	23.0	23.1	0.1	24.2	24.8	0.6	29.8	29.0	0.8
Phosphoform 5	24.0	24.0	0.0	24.0	24.0	0.0	25.1	25.0	0.1	26.9	27.0	0.1

Fig. 15. The *Pho4* multisite phosphorylation model validation. The table represents the quantification of site preferences calculated as the amount of each phosphopeptide as the percentage of the total phosphorylation for every phosphoform. The *bio* column is extracted from [9] (Figure 3(g)) derived from *in vitro* experiments, the sπ column contains the results of our *in silico* simulations and the *err* column contains the absolute differences between the two approaches.

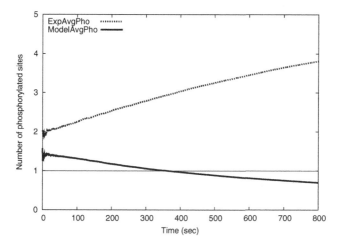

Fig. 16. The *Pho4* multisite phosphorylation simulation. Here are represented the modifications during the reaction of the two different ways to quantify the number of phosphorylations for binding event. *ExpAvgPho* represents the calculation without considering the bindings with no phosphorylations (as performed in [9]), *ModelAvgPho* the quantification obtained from the stochastic π-calculus model considering also the null events. The horizontal line points out the limit under which the phosphorylation is considered completely distributive.

that were already phosphorylated and permits kinase-substrate unbindings before any phosphorylation events thus allowing the possibility of binding events without phosphorylations. Dividing the total number of binding events for the total number of phosphorylation events the model permits to estimate the average number of phosphorylations per binding in a correct, complete and natural way (we call this calculation *ModelAvgPho*). Applying statistical methods to repeated stochastic *in silico* experiments referred to a small time window at the beginning of the reaction in order to improve robustness and precision, we obtained a value of the average number of phosphorylations for binding event of 1.45 which is 30% lower than the estimation without non effective bindings (2.06). It means that the *Pho80-Pho85* kinase activity on *Pho4* substrate is much closer to a distributive behaviour than to a processive one and that the kinetics of the *Pho4* regulation in phosphate starvation depends more on the number of kinase-substrate bindings than on the multiple phosphorylations per binding event with respect to the findings reported in [9]. Moreover, moving the time window along the reaction course, it is possible to notice, as shown in Figure 16, that the kinase begins to have a complete distributive behaviour (the average phosphorylation value is lower than 1) after about 6 minutes. Theoretically, as confirmed by Figure 17 which extends the duration time of the analysis in Figure 16, the model based phosphorylation value (*ModelAvgPho*) correctly approaches zero as the reaction goes to completion while the quantification following the calculation used in [9] (*ExpAvgPho*) reaches 5

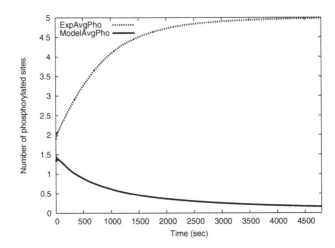

Fig. 17. The *Pho4* multisite phosphorylation simulation. This plot represents the *ExpAvgPho* and *ModelAvgPho* as defined in Figure 16 but extending the simulation to 80 minutes in order to show how *ExpAvgPho* approaches 5 and *ModelAvgPho* approaches 0 as the reaction goes to completion.

when all the *Pho4* proteins are phosphorylated and only phosphoforms 5 are present.

6.3   Kinetic Analysis of *Pho4* Phosphorylation Profiles Predicts Biological Outcomes

The *Pho4* multisite phosphorylation computational model allows the analysis of the concentration of each of the 32 phosphorylation profiles among the reaction course it shows that only few species (one or at most three) per phosphoform are predominant (see Figure 18 for the analysis of each phosphoform and Figure 19 where only the predominant profiles are reported).

Figure 20 reports the number of phosphorylations on the five different sites and on s_2 and s_3 at the same time, thus providing for each time instant the percentage of *Pho4* molecules having particular functions and behaviours. For example, it states that half of the transcription factors are phosphorylated on s_6 after about three minutes and so inactivated for the cooperative binding with *Pho2* to the promoter of *Pho5* gene, while about ten minutes are necessary to have half of the promoters phosphorylated both on s_2 and s_3 and so ready to be exported in the cytoplasm by the extranuclear transporter *Msn5*. The simulation confirms that the *Pho5* transcription inhibition is the result of two different regulations (as stated in [36]): the cooperative inhibition and the extranuclear localization that act simultaneously but with different dynamics. More precisely we give an account of the fact that the regulations of *Pho2*-dependent and *Pho2*-independent genes are separated and influenced by two different phosphorylation set of sites. The s_2 and s_3 sites regulate the nuclear localization of the *Pho4* and

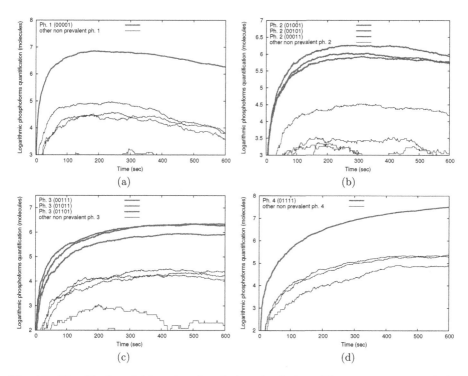

Fig. 18. The *Pho4* multisite phosphorylation simulation. The representation of the absolute logarithmic amount (in number of molecules) of every phosphorylation pattern of phosphoform 1 (a), 2 (b), 3 (c) and 4 (d). The total amount of *Pho4* proteins is 10000. For each phosphoform the predominant species are highlighted (in red) and identified with five binary values representing respectively the phosphorylation state of s_1, s_2, s_3, s_4 and s_6 (1 denotes the phosphorylation). Some non predominant species do not appear in the plots because their expression is too low.

so its possibility to control the *Pho2*-independent genes, while s_6 inhibits the cooperative binding with *Pho4* and so the *Pho2*-dependent gene expression. Moreover, the very low number of *Pho4* phosphorylated on s_2 and on s_3 but not on s_6 (always under 0.6% of total *Pho4*) means that the *Pho2*-dependent gene regulation is not influenced by the nuclear exportation. These predictions allow us to state some new hypotheses about the *Pho2* role. In particular it follows that s_6 is the only site that regulates the *Pho2*-dependent genes because it is very unlikely (less than 1% of the cases) that *Pho4* proteins not phosphorylated on s_6 are exported out of the nucleus. From the kinetics point of view, with the results of the *in silico* analyses, we can conclude that the s_6 phosphorylation dynamic determines the *Pho2*-dependent genes transcription kinetics while the s_2 and s_3 ones determine the *Pho2*-independent genes transcription kinetics.

An observation that confirms an already noticed behaviour [9,36] is that the s_4 is phosphorylated rapidly with respect to s_2 and s_3 especially if considered

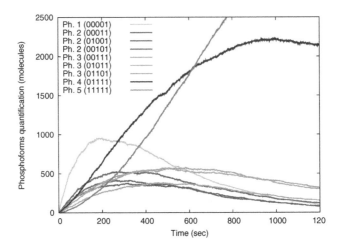

Fig. 19. The *Pho4* multisite phosphorylation simulation. This graph represents the absolute amount during the reaction course of the prevalent phosphorylation patterns of every phosphoform (the total number of *Pho4* molecules is 10000). In particular every phosphoform has only one or three (in the case of phosphoform 2 and 3) predominant patterns. The species are identified with five binary values representing respectively the phosphorylation state of s_1, s_2, s_3, s_4 and s_6 (1 denotes the phosphorylation).

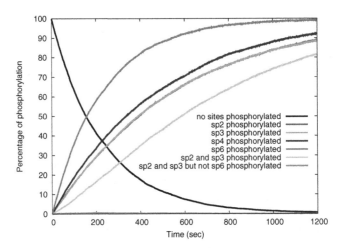

Fig. 20. The *Pho4* multisite phosphorylation simulation. This plot represents the percentage of *Pho4* species phosphorylated in different and possible combined sites (s_1 site is not considered because its role is unknown). The y axis represents the percentages with respect to the total mount of *Pho4* (10000 molecules), the x axis the time expressed in seconds.

together, meaning that the nuclear reimportation is prevalently inhibited before forcing the cytoplasm localization and thus avoiding a futile iterative cycle.

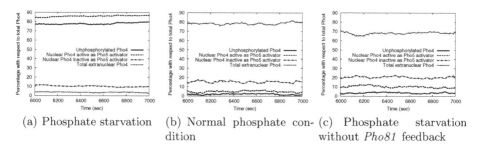

(a) Phosphate starvation   (b) Normal phosphate condition   (c) Phosphate starvation without *Pho81* feedback

Fig. 21. The whole PHO pathway simulation. The three plots show the behaviour of the *Pho4* during phosphate starvation (a), normal phosphate condition (b), and starvation condition without the *Pho81* feedback loop in the model (c). The time is measured from the beginning of the reaction.

6.4   PHO Pathway Results

No precise quantitative biological information is available about the entire PHO pathway but the *Pho4* regulation characteristics and so the *in silico* predictions have a lower confidence with respect to the ones performed with the *Pho4* multisite phosphorylation model; however, the model proves to be consistent with the qualitative description of different phosphate conditions.

Considering the level of unphosphorylated *Pho4*, of nuclear *Pho4* active and inactive as *Pho5* promoter activator and of extranuclear *Pho4*, we show computationally the evidence of the crucial role of the *Pho81* feedback loop. This happens because the removal from the model of the *Pho81* feedback influence causes the *in silico* experiments to exhibit a starvation induced behaviour (Figure 21(c)) more similar to the normal phosphate condition (Figure 21(b)) than to the starvation condition with the feedback mechanism (Figure 21(a)).

The analysis of the nuclear *Pho4* confirms quantitatively the existence of partial phosphate starvation responses to intermediate levels of phosphate in the medium as stated in [36]. In Figure 22(a) we can notice that for percentages between 10% and 20% the *Pho4* is prevalently nuclear but the cooperative binding with *Pho2* is inhibited allowing only the transcription of the *Pho2* independent genes. This is reasonably due to the already discussed different phosphorylation kinetics of s_6 site and s_2 and s_3 sites that cause, as the percentage of active *Pho81* increases, an initial fast increasing of nuclear inactive *Pho4* concentration (for the rapid s_6 phosphorylation) followed by a gradual decreasing of the same species (for the slower combined phosphorylation of s_2 and s_3).

We performed a series of computational analyses about the expression levels of *Pho5*, a phosphatase responsible for example of the *Pho84* regulation [39], with respect to increasing phosphate concentrations. Since the phosphate concentration in the medium cannot be directly included in the π-calculus formalism, the input of the PHO pathway model is indirectly defined by the initial percentage of *Pho81* proteins active as inhibitors. We can notice from the simulations (Figure 22(b)) that a low percentage of active *Pho81* is sufficient to increase the

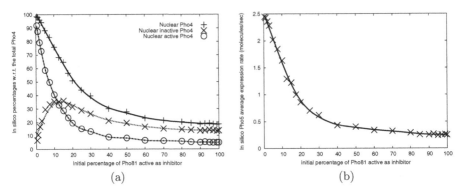

Fig. 22. The whole PHO pathway simulations in different phosphate conditions. The graph in (a) describes the predicted levels of nuclear *Pho4* and of transcriptional active and inactive nuclear *Pho4* with respect to the initial percentage of *Pho81* active as *Pho4* inhibitor. The graph in (b) represents the model-based *in silico* estimations of the *Pho5* expression rate with different initial percentages of *Pho81* active as *Pho4* inhibitor. The *in silico* levels are interpolated to highlight the overall trends.

phosphatase transcription, but it is required that at least 50% of *Pho81* proteins are active to have the complete starvation signal. In fact for percentages greater than 50% the *Pho5* expression is more or less stable. Assuming to have experimental values of the *Pho5* expression in different phosphate conditions, through this prediction it is possible to characterize the dynamics of the *Pho81 minimum domain* sensing of phosphate conditions [40] which is still almost unknown.

Finally we checked the qualitative consistency of the PHO pathway model with microarray data. The expression level ratios of *Pho5* and *Pho81* between normal and phosphate starvation conditions agree qualitatively with the microarray dataset available in [41] in the sense that the computational model is able to predict if they are overexpressed or underexpressed.

7   Analysis of Process Calculi Features

7.1   *In silico* Process Calculi for Multisite Phosphorylations

The proposed *in silico* approach to the *Pho4* multisite phosphorylation mechanism revealed the possibility to detect, explore and quantify some aspects that have not been analysed previously mainly because of the intrinsic difficulty in planning proper experimental or computational approaches. This property is highlighted by the model based estimation of the average number of *Pho4* phosphorylation per *Pho80-Pho85* binding event.

The model proved to be able to handle the complexity of the phosphorylation mechanism and to be consistent with biological data. With such a model it is possible to computationally quantify every possible event. Also in relation to aspects already known in literature the *in silico* experiments adds to the qualitative knowledge also an in depth quantitative description permitting new possible analyses and predictions. Another desirable feature of the model is the time-continuous property of the *in silico* analyses typically not peculiar of the *in vitro* or *in vivo* ones.

The multisite phosphorylation stochastic π-calculus model can be applied to all the phosphorylation mechanisms with a desired number of phosphorylation sites. Considering that multisite phosphorylation is one of the major cellular regulation mechanisms, the number of cases in which the proposed model can be applied for *in silico* experiments is potentially very large. Among all the possible multisite phosphorylation cases the most potentially promising ones are those that are still partially unknown from a biological point of view but for which indirect experimental values sufficient for the tuning and validation are available. The values of the model to set are the number of phosphorylation sites, the affinity between the kinase and the substrate, the phosphorylation rate of each site and the kinase-substrate disassociation rate. With respect to ODE based approach, no *a priori* assumptions on the general behaviour of the system are needed, no convergence and stability problems can occur even for small quantities, the non-determinism and noise are natively included in the simulator and a single simulation can investigate every aspect of the reaction.

Note that we can also use the model to try to infer the rates. We can artificially set the rates and see then how they affect the overall behaviour. This approach can be useful to plan specific *in vitro* or *in vivo* experiments to validate or disprove hypotheses.

7.2   Biological Modelling of Discrete and Null Events

Among the possible paradigms to the *in silico* biological approach, the process calculi have formal, computational and modelling advantages as reviewed in [42]. In this work we found another desirable feature related to the biological potentialities of process calculi. It regards the possible description of biological behaviours that are intrinsically very hard to measure through the qualitative knowledge of the mechanism and the indirect quantitative traditional experimental data. The binding and subsequent unbinding of the kinase to the substrate without phosphorylations is hardly detectable with *in vivo* or *in vitro* techniques because its duration is very limited and does not change any species concentration; on the other hand, modelling approaches based on continuous time and concentration assumption, like ODE, cannot easily handle these cases. The process calculi description, instead, can detect and quantify every possible single or multiple discrete transition step of the system. It follows that the central problem for process calculi in this context is not the quantification and simulation phase, but the adequate setting of the model parameters with respect to measured values; the tuning and validation phase cannot always be successfully performed and it heavily depends on the model structure and on the type and precision of the indirect biological data. In the specific case of the multisite phosphorylation of *Pho4* the rates can be set properly using only the *in vitro* phosphoform quantification and the site preferences.

Summing up, the innovative theoretical biological modelling feature of the process algebras highlighted by this work consists in the possibility, under certain conditions about the model structure and the available indirect data, to systematically describe biological events that are null (in the sense that no concentration variation are appreciable), discrete and hardly measurable.

7.3   The Problem of Explosion of the Number of Process Definition

The conceptual model of the multisite phosphorylation needs 2^n states where n is the number of phosphorylation sites, to represent each possible substrate phosphorylation profile. In our case the states are $2 \times 2^n = 2^{n+1}$ because the substrate and the kinase-substrate complex are conceptually separated. In the proposed π-calculus model there is a different process definition for each different state. This means that as the number of phosphorylation sites become larger, the number of the needed process definitions grows exponentially leading to a huge and non-readable model in addition to the computational and memory simulation problems.

The SPi machine, and so the SPiM simulator (see Appendix A), permits the reduction of the processes with respect to the states allowing the parameters passing between the processes. In the phosphorylation model the idea could be to use only two processes (one for the substrate and one for the kinase-substrate complex) denoting the states with $\lceil \log_2(n) \rceil$ boolean values (where n is the number of states) that are passed and possibly modified every time a recursive call of the processes is performed. Obviously every operation of the two processes that depends on the number of phosphorylation sites must be "guarded" by an *if-then-else* like operator which permits to enable only the desired operations. Different semantics of the *if-then-else* are possible (for example blocking or non-blocking behaviours) and this heavily influences the development of the model. However, the most important theoretical problem is the semantics of the non-deterministic choice (Appendix A); in fact it states that when a branch is selected the other ones cannot be executed. So it is clear that if the *if-then-else* like operator is executed, the relative non-deterministic choice branch is selected independently if the conditional statement evaluates to *true* or to *false*.

Theoretically, the expressivity of the π-calculus does not decrease, under certain conditions, if the non-deterministic choice is removed from the language [43]. However, from a modelling point of view, the non-deterministic choice is a powerful, intuitive and elegant primitive and with the parallel composition is one of the main reasons why process calculi are particularly suitable in biology. It is important to underline, in fact, that the intuitiveness of the modelling approach is a key factor for its success among the biologist community.

Summing up, in the π-calculus the process definition reduction is intrinsically prevented by the use of the non-deterministic choice. The way for an effective reduction of the definitions can be represented, rather then by complex and possibly useless model modifications, by other π-calculus based languages like *Beta Binders* [44,45] that natively allows the multiple state handling by means of particular abstractions called interfaces.

7.4 Compositionality for Handling Systems Biology Complexity

The modelling approach used in this work consists in the development of small models regarding single biological aspects that can be integrated to specify the entire studied pathway. The PHO pathway modelling by means of a set of general and distinct sub-models with very few code adaptations shows the theoretical crucial role of the compositionality property [42] of the process algebras that is a modelling feature not handled by ODE based techniques and which can reasonably be the key for handling the intrinsic complexity of systems biology.

The proposed multisite phosphorylation, promoter activity, transmembrane transporting, synthesis and degradation models are recurrent aspects in all the pathways. Starting from these models it is possible to implement a sort of "computational biological sub-paths library"; the idea is then to adapt the predefined general small models of the library to specific biological entities and mechanisms and aggregate them by means of compositionality to model complex pathway in a relatively easy and fast way. Applying compositionality also to the modelled pathways we can really obtain a powerful framework to lift the computational biology to the systems biology level.

8 Conclusions

In this work we developed stochastic π-calculus models for some biological recurrent mechanisms that, by means of compositionality, were adapted for describing the *Pho4* multisite phosphorylation and the PHO pathway in *Saccharomyces cerevisiae*. The *Pho4* multisite phosphorylation model was tuned to be consistent with available *in vitro* biological data and validated with good precision. The stochastic *in silico* predictions we performed suggest that the *Pho80-Pho85* phosphorylation activity on the *Pho4* substrate should be considered distributive and not processive or semi-processive as previously found, estimate quantitatively the concentration variations of all possible phosphorylation profiles, and state that the *Pho2*-dependent and independent genes are regulated by different set of phosphorylation sites which also have distinct kinetic roles. The PHO pathway model proves to be consistent with the biological observations even if precise data is not available for all the sub-paths. However, we computationally confirm the crucial role of the *Pho81* feedback loop and the existence of intermediate responses reflecting partial phosphate starvation conditions. We also propose an analysis regarding the starvation gene expression with respect to the percentage of active *Pho81* inhibitors. In the work we discuss the potentialities of the general multisite phosphorylation model and of the process algebras *in silico* approach. In particular a new advantage of process algebras for computational biology is highlighted and it regards the simulation of discrete and null biological event that are intrinsically hard to measure.

The concrete biological predictions detailed in the work represent, as far as we know, the first significant biological results achieved with process algebras methods.

Acknowledgment

The authors would like to thank Corrado Priami of The Microsoft Research - University of Trento Centre for Computational and Systems Biology for insightful comments on the manuscript. This work was partially supported by FIRB - *Computational tools for Systems Biology* and by BISCA - *Sistemi e calcoli di ispirazione biologica e loro applicazioni.*

References

1. Seet, B., Dikic, I., Zhou, M., Pawson, T., et al.: Reading protein modifications with interaction domains. Nat. Rev. Mol. Cell. Biol. 7, 473–483 (2006)
2. Yang, X.: Multisite protein modification and intramolecular signaling. Oncogene 24, 1653–1662 (2005)
3. Seo, J., Lee, K.: Post-translational modifications and their biological functions: proteomic analysis and systematic approaches. J. Biochem. Mol. Biol. 37(1), 35–44 (2004)
4. Gunawardena, J.: Multisite protein phosphorylation makes a good threshold but can be a poor switch. Proc. Natl. Acad. Sci. USA 102(41), 14617–14622 (2005)
5. Holmberg, C., Tran, S., Eriksson, J., Sistonen, L.: Multisite phosphorylation provides sophisticated regulation of transcription factors. Trends Biochem. Sci. 27(12), 619–627 (2002)
6. Cohen, P.: The regulation of protein function by multisite phosphorylation–a 25 year update. Trends Biochem. Sci. 25(12), 596–601 (2000)
7. Mayya, V., Rezual, K., Wu, L., Fong, M., Han, D.: Absolute Quantification of Multisite Phosphorylation by Selective Reaction Monitoring Mass Spectrometry: Determination of Inhibitory Phosphorylation Status of Cyclin-Dependent Kinases. Mol. Cell. Proteomics 5(6), 1146 (2006)
8. Glinski, M., Weckwerth, W.: Differential Multisite Phosphorylation of the Trehalose-6-phosphate Synthase Gene Family in Arabidopsis thaliana: A Mass Spectrometry-based Process for Multiparallel Peptide Library Phosphorylation Analysis. Mol. Cell. Proteomics 4(10), 1614–1625 (2005)
9. Jeffery, D., Springer, M., King, D., O'Shea, E.: Multi-site phosphorylation of Pho4 by the cyclin-CDK Pho80-Pho85 is semi-processive with site preference. J. Mol. Biol. 306(5), 997–1010 (2001)
10. Markevich, N.I., Hoek, J.B., Kholodenko, B.N.: Signaling switches and bistability arising from multisite phosphorylation in protein kinase cascades. J. Cell. Biol. 164(3), 353–359 (2004)
11. Batchelor, E., Goulian, M.: Robustness and the cycle of phosphorylation and dephosphorylation in a two-component regulatory system. Proc. Natl. Acad. Sci. USA 100(2), 691–696 (2003)
12. Huang, C., Ferrell Jr, J.: Ultrasensitivity in the mitogen-activated protein kinase cascade. Proc. Natl. Acad. Sci. USA 93, 10078–10083 (1996)
13. Cateau, H., Tanaka, S.: Kinetic analysis of multisite phosphorylation using analytic solutions to Michaelis-Menten equations. J. Theor. Biol. 217(1), 1–14 (2002)
14. Regev, A., Silverman, W., Shapiro, E.: Representation and simulation of biochemical processes using the pi-calculus process algebra. Pac. Symp. Biocomput. 459, 70 (2001)

15. Priami, C., Regev, A., Shapiro, E., Silverman, W.: Application of a stochastic name-passing calculus to representation and simulation of molecular processes. Inform. Process. Lett. 80(1), 25–31 (2001)

16. Errampalli, D., Priami, C., Quaglia, P.: A Formal Language for Computational Systems Biology. OMICS 8(4), 370–380 (2004)

17. Ciocchetta, F., Priami, C., Quaglia, P.: Modeling Kohn Interaction Maps with Beta-Binders: An Example. T. Comp. Sys. Biol. 3 (2005)

18. Lecca, P., Priami, C., Laudanna, C., Constantin, G.: A BioSpi model of lymphocyte-endothelial interactions in inflamed brain venules. Pac. Symp. Biocomput. 521, 32 (2004)

19. Kuttler, C.: Simulating bacterial transcription and translation in a stochastic pi calculus. T. Comp. Sys. Biol. 4220, 113–149 (2006)

20. Curti, M., Degano, P., Priami, C., Baldari, C.: Modelling biochemical pathways through enhanced π-calculus. Theor. Comput. Sci. 325(1), 111–140 (2004)

21. Wykoff, D., O'Shea, E.: Phosphate Transport and Sensing in Saccharomyces cerevisiae. Genetics 159(4), 1491–1499 (2001)

22. Phillips, A.: The stochastic Pi machine (SPiM),
    http://research.microsoft.com/~aphillip/spim/

23. Gillespie, D.: Exact stochastic simulation of coupled chemical reactions. J. Phys. Chem. 81(25), 2340–2361 (1977)

24. Gregory, P., Barbari, S., Hörz, W.: Transcriptional Control of Phosphate-regulated Genes in Yeast: the Role of Specific Transcription Factors and Chromatin Remodeling Complexes in vivo. Food Technol. Biotechnol. 38, 295–303 (2000)

25. Persson, B., Lagerstedt, J., Pratt, J., Pattison-Granberg, J., Lundh, K., Shokrollahzadeh, S., Lundh, F.: Regulation of phosphate acquisition in Saccharomyces cerevisiae. Curr. Genet. 43(4), 225–244 (2003)

26. Waters, N., Knight, J., Creasy, C., Bergman, L.: The yeast Pho80–Pho85 cyclin–CDK complex has multiple substrates. Curr. Genet. 46(1), 1–9 (2004)

27. Carroll, A., O'Shea, E.: Pho85 and signaling environmental conditions. Trends Biochem. Sci. 27, 87–93 (2002)

28. Komeili, A., O'Shea, E.: Roles of Phosphorylation Sites in Regulating Activity of the Transcription Factor Pho4. Science 284(5416), 977 (1999)

29. Byrne, M., Miller, N., Springer, M., O'Shea, E.: A distal, high-affinity binding site on the cyclin-CDK substrate Pho4 is important for its phosphorylation and regulation. J. Mol. Biol. 335(1), 57–70 (2004)

30. Bhoite, L., Allen, J., Garcia, E., Thomas, L., Gregory, I., Voth, W., Whelihan, K., Rolfes, R., Stillman, D.: Mutations in the pho2 (bas2) transcription factor that differentially affect activation with its partner proteins bas1, pho4, and swi5. J. Biol. Chem. 277(40), 37612–37618 (2002)

31. Rudolph, H., Hinnen, A.: The yeast PHO5 promoter: phosphate-control elements and sequences mediating mRNA start-site selection. Proc. Natl. Acad. Sci. USA 84(5), 1340–1344 (1987)

32. Barbaric, S., Munsterkotter, M., Goding, C., Horz, W.: Cooperative Pho2-Pho4 interactions at the PHO5 promoter are critical for binding of Pho4 to UASp1 and for efficient transactivation by Pho4 at UASp2. Mol. Cell. Biol. 18(5), 2629–2639 (1998)

33. Barbaric, S., Munsterkotter, M., Svaren, J., Horz, W.: The homeodomain protein Pho2 and the basic-helix-loop-helix protein Pho4 bind DNA cooperatively at the yeast PHO5 promoter. Nucleic Acids Res. 24(22), 4479–4486 (1996)

34. Kaffman, A., Rank, N., O'Shea, E.: Phosphorylation regulates association of the transcription factor Pho4 with its import receptor Pse1/Kap121. Genes. Dev. 12(17), 2673–2683 (1998)
35. Kaffman, A., Rank, N., O'Neill, E., Huang, L., O'Shea, E.: The receptor Msn5 exports the phosphorylated transcription factor Pho4 out of the nucleus. Nature 396(6710), 482–486 (1998)
36. Springer, M., Wykoff, D., Miller, N., O'Shea, E.: Partially phosphorylated pho4 activates transcription of a subset of phosphate-responsive genes. PLoS Biol. 1, 2 (2003)
37. Ghaemmaghami, S., Huh, W., Bower, K., Howson, R., Belle, A., Dephoure, N., O'Shea, E., Weissman, J.: Global analysis of protein expression in yeast. Nature 425(6959), 737–741 (2003)
38. Huh, W., Falvo, J., Gerke, L., Carroll, A., Howson, R., Weissman, J., O'Shea, E.: Global analysis of protein localization in budding yeast. Nature 425(6959), 686–691 (2003)
39. Martinez, P., Zvyagilskaya, R., Allard, P., Persson, B.: Physiological regulation of the derepressible phosphate transporter in Saccharomyces cerevisiae. J. Bacteriol. 180(8), 2253–2256 (1998)
40. Swinnen, E., Rosseels, J., Winderickx, J.: The minimum domain of Pho81 is not sufficient to control the Pho85-Rim15 effector branch involved in phosphate starvation-induced stress responses. Curr. Genet. (2005)
41. Ogawa, N., DeRisi, J., Brown, P.: New Components of a System for Phosphate Accumulation and Polyphosphate Metabolism in Saccharomyces cerevisiae Revealed by Genomic Expression Analysis. Mol. Biol. Cell. 11(12), 4309–4321 (2000)
42. Priami, C., Quaglia, P.: Modelling the dynamics of biosystems. Brief Bioinform. 5(3), 259–269 (2004)
43. Nestmann, U., Pierce, B.: Decoding Choice Encodings. Information and Computation 163(1), 1–59 (2000)
44. Priami, C., Quaglia, P.: Beta binders for biological interactions. In: Danos, V., Schachter, V. (eds.) CMSB 2004. LNCS (LNBI), vol. 3082, pp. 20–33. Springer, Heidelberg (2005)
45. Degano, P., Prandi, D., Priami, C., Quaglia, P.: Beta-binders for biological quantitative experiments. Proceedings of QAPL 2006 164(3), 101–117 (2006)
46. Priami, C.: Stochastic π-Calculus. The Computer Journal 38(7), 578 (1995)
47. Bergstra, J., Ponse, A., Smolka, S.: Handbook of Process Algebra. Elsevier Science Inc, New York (2001)
48. Phillips, A., Cardelli, L.: A correct abstract machine for the stochastic π-calculus. ENTCS. Elsevier, Amsterdam (2005)

A   The Stochastic π-Calculus and SPiM

The stochastic π-calculus was proposed by [46] as an extension of the π-calculus, a process algebra originally developed for describing concurrent computations whose configuration may change during the execution [47]. Central to the calculus are the names because they play a dual role of communication channels and variables. The basic example regards the transfer of a name among a channel between two processes running in parallel; the receiver can use the name as a channel for further dynamic interactions with other processes in the system.

Table 1. The syntax of the π-calculus

Prefixes

$$\pi ::= \tau \mid x\langle y\rangle \mid x(y)$$

Processes

$$P ::= \mathbf{0} \mid (\pi, r).P \mid P + P \mid P|P \mid !P \mid \nu x\, P$$

Table 2. The free and the bound names of the stochastic π-calculus

P	$\mathsf{n}(P)$	$\mathsf{bn}(P)$	$\mathsf{fn}(P)$
$\mathbf{0}$	$\emptyset$	$\emptyset$	$\emptyset$
$(\tau.r, R)$	$\mathsf{n}(R)$	$\mathsf{bn}(R)$	$\mathsf{fn}(R)$
$(x\langle y\rangle, r).R$	$\{x, y\} \cup \mathsf{n}(R)$	$\mathsf{bn}(R)$	$\{x, y\} \cup \mathsf{fn}(R)$
$(x(y), r).R$	$\{x, y\} \cup \mathsf{n}(R)$	$\{y\} \cup \mathsf{bn}(R)$	$\{x\} \cup \mathsf{fn}(R)$
$R\|Q$	$\mathsf{n}(R) \cup \mathsf{n}(Q)$	$\mathsf{bn}(R) \cup \mathsf{bn}(Q)$	$\mathsf{fn}(R) \cup \mathsf{fn}(Q)$
$R + Q$	$\mathsf{n}(R) \cup \mathsf{n}(Q)$	$\mathsf{bn}(R) \cup \mathsf{bn}(Q)$	$\mathsf{fn}(R) \cup \mathsf{fn}(Q)$
$!R$	$\mathsf{n}(R)$	$\mathsf{bn}(R)$	$\mathsf{fn}(R)$
$\nu x\, R$	$\{x\} \cup \mathsf{n}(R)$	$\{x\} \cup \mathsf{bn}(R)$	$\mathsf{fn}(R) \setminus \{x\}$

The stochastic version integrates in the formalism the exponential probability distribution in order to quantitatively accommodate the times at which the communications occur and the system evolves.

The syntax of the stochastic π-calculus, formally shown in Table 1, consists in three atomic actions, called prefixes, and in processes. The prefixes are the output $x\langle y\rangle$ of name y on a channel x, the input $x(y)$ on a channel x in which y acts as the placeholder for the input, and the silent action τ which denotes an action invisible to an external observer of the system. Each prefix is associated in the processes definition with a stochastic rate r, which is the parameter of the exponential distribution describing the duration of the activity; the intuition behind a process guarded by a prefix $(\pi, r).P$ is that process P is enabled only after the performing of the action π whose duration depends on the stochastic rate r. The simplest process is the null process or deadlock $\mathbf{0}$, which can do nothing. The non-deterministic choice $P + P$ permits the exclusive selection between two processes, while the parallel composition $P|P$ denotes the combined behavior of processes executing in parallel. The replication $!P$ allows an unbounded number of copies of the same process in parallel composition. The restriction operator $\nu x\, P$ specifies the scope of a name binding statically the name x in P.

The names in the calculus can be free or bound. Intuitively, the restriction and the input prefix are the only processes that bind names; $\nu x\, P$ binds x in P, $(x(y), r).Q$ binds y in Q. Table 2 formally presents the recursive definition of $\mathsf{n}(P)$ (the names of a process P), of $\mathsf{bn}(P)$ (the names occurring bound in P) and of $\mathsf{fn}(P)$ (the names with a not bound - i.e. free - occurrence in P), with respect to the syntactic structure of P.

The syntactic rules are in some sense too rich, allowing multiple definitions for the same intuitive behaviour (e.g. $P|Q$ and $Q|P$ for the parallel composition of P and Q which is intuitively commutative). We thus need the definition of

Table 3. Rules of structural congruence for the π-calculus

1. If P and Q are identical under α-conversions then $P \equiv Q$
2. $(\mathcal{P}/_\equiv, |, \mathbf{0})$ and $(\mathcal{P}/_\equiv, +, \mathbf{0})$ are commutative monoids
3. The unfolding law: $\,! P \equiv P \,|\, ! P$
4. Scope restriction laws:

$$\nu x\, \mathbf{0} \equiv \mathbf{0}$$
$$\nu x\,(P \mid Q) \equiv P \mid \nu x\, Q \quad \text{if } x \notin \mathsf{fn}(P)$$
$$\nu x\,(P + Q) \equiv P + \nu x\, Q \quad \text{if } x \notin \mathsf{fn}(P)$$
$$\nu x\, \nu y\, P \equiv \nu y\, \nu x\, P$$

Table 4. Operational semantics of the π-calculus

$$(\text{prefix}) \quad \frac{}{(\alpha, r).\, P \xrightarrow{(\alpha, r)} P}$$

$$(\text{com}) \quad \frac{P \xrightarrow{(x(z), r)} P',\ Q \xrightarrow{(x\langle y\rangle, r)} Q'}{P \mid Q \xrightarrow{(\tau, r)} P'\{y/z\} \mid Q'}$$

$$(\text{par}) \quad \frac{P \xrightarrow{(\alpha, r)} P'}{P \mid Q \xrightarrow{(\alpha, r)} P' \mid Q} \quad \text{if } \mathsf{bn}(\alpha) \cap \mathsf{fn}(Q) = \emptyset$$

$$(\text{sum}) \quad \frac{P \xrightarrow{(\alpha, r)} P'}{P + Q \xrightarrow{(\alpha, r)} P'}$$

$$(\text{res}) \quad \frac{P \xrightarrow{(\alpha, r)} P'}{\nu x\, P \xrightarrow{(\alpha, r)} \nu x\, P'} \quad \text{if } x \notin \mathsf{n}(\alpha)$$

$$(\text{struct}) \quad \frac{P' \equiv P,\ P \xrightarrow{(\alpha, r)} Q,\ Q \equiv Q'}{P' \xrightarrow{(\alpha, r)} Q'}$$

structural congruence which is the identity relation up to the processes structure presented in Table 3, where α-conversion is the substitution of one or more bound names in a process P with fresh names (i.e. with names that does not cause conflicts) and $\mathcal{P}/_\equiv$ is the set of classes of congruence among processes.

The semantics of the calculus is given through the classical SOS (structural operational semantics) approach, which gives rise to a labelled transition system. The transitions are of the kind $P \xrightarrow{(\alpha, r)} Q$ for some set of actions ranged over by α and rates r. The class of actions, ranged over by the meta-variable α, are the internal action τ, the output action $x\langle y\rangle$ and the input action $x(y)$. The free and the bound names of the actions are the same as the ones defined for the prefixes in Table 2.

We can now introduce the operational semantics briefly explaining its reduction rules shown in Table 4. The reduction rule **(prefix)** is applied on a prefix that is consumed according with its stochastic rate r, after which the execution proceeds with the guarded process; **(com)** captures the synchronous communication capability between two parallel processes that are willing to perform an input and an output on the same channel and with the same rate r and states that, once the communication occurs with a rate r, the prefixes are consumed and the receiver substitutes the free occurrences of the placeholder with the received name; **(par)** regulates the evolution of a process in a parallel composition structure that consumes a prefix (which must not have a correspondence in the other parallel processes) with a rate r and that can thus

1 $\nu\, ioniz\_ch$
2 $!(Na(), r_\infty). \nu\, e^- (ioniz\_ch\langle e^-\rangle, r_{ion}). (Na^+\langle\rangle, r_\infty). \mathbf{0}$
3 $|\, !(Cl(), r_\infty). (ioniz\_ch(e^-), r_{ion}). (Cl^-\langle\rangle, r_\infty). \mathbf{0}$

Fig. 23. The model of the ionization of Na

evolve maintaining the same context and rate in the parallel composition; **(sum)** states that only one process in a non-deterministic choice can evolve and, once it evolves, any other process in the summation is eliminated; **(res)** permits a process with restriction to evolve with the same rate maintaining the restriction if the restricted name is not contained in the consumed prefix; finally **(struct)** formalizes the idea that processes that are structurally congruent have the same behaviour.

A small example. The model in Figure 23 represents the stochastic π-calculus description of Na cationization by Cl $(Na + Cl \Rightarrow Na^+ + Cl^-)$. We use this simple example to introduce some abstractions used in describing the PHO pathway models. The Na process (line 2) can be activated with an *empty communication* (i.e. a communication without passing names and thus used for synchronization only) on the Na channel with infinite rate (the stochastic rate of the prefix is r_∞, so that the synchronization can happen as soon as it becomes available) and similarly for the Cl process (line 3). The ionization capability is modelled with a specific channel ($ioniz\_ch$), that is declared outside the scope of the interested processes and for this reason is said *global*. The ionization event occurs applying the **(com)** rule to the $(ioniz\_ch\langle e^-\rangle, r_{ion})$ prefix of Na and on the $(ioniz\_ch(e^-), r_{ion})$ prefix of Cl passing the name e^-; note that all the hypotheses of the rule are respected since the rate of input and output operations is the same (r_{ion}) and both processes can evolve. The transmitted name e^- representing the electron is declared with a restriction internal to the Na process, meaning that e^- is a *private* or *restricted* name in the Na process. These abstractions allow us to say that Na cationization is modelled with a synchronous communication on a global channel of a private name.

The SPiM simulator. The Stochastic Pi Machine (SPiM) [48] is a simulator of the stochastic π-calculus (SPi) that can be used to simulate models of biological systems. SPi uses a variant of the stochastic π-calculus in which the replication and the non-deterministic choice are guarded (i.e. the replicated processes and the branches of the non-deterministic choice start with a prefix). Moreover, in SPi, each channel x is associated with a corresponding reaction rate given by $rate(x)$ avoiding to include directly the rates in the prefixes definition. SPiM executes a process P using a stochastic selection procedure based on the Gillespie algorithm [23] in order to determine which is the next reaction channel. In particular, the selection considers also the notion of channel activity, defined as the number of possible combinations of inputs and outputs on channel x in a process P. SPiM is available for download [22].

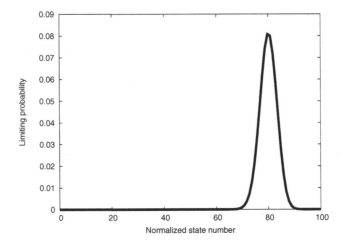

Fig. 24. The limiting probability distribution of the CTMC with 150 states

B   The Model for the Dynamic Regulation of the Number of Processes as a Continuous-Time Markov Chain

The model presented in 3.2 and shown in Figure 6, developed for regulating the relative amount of a certain set of processes in a state S_1 with respect to another set of processes in state S_2, can be seen as a continuous-time Markov chain (CTMC) of a birth-death process. The finite chain with n states is defined by the following elements of the infinitesimal generator matrix:

$$q_{ij} = \begin{cases} (n-i)\,r_1 & 0 \le i \le n-1,\ j = i+1 \\ i\,r_2 & 1 \le i \le n,\ j = i-1 \\ (i-n)\,r_1 - i\,r_2 & 0 \le i \le n,\ i = j \\ 0 & \text{otherwise} \end{cases}$$

where r_1 and r_2 are the changing rate between the two states. We derived the limiting probabilities (denoting r_1/r_2 with ρ):

$$\pi_i = \binom{n}{i}\rho^i(\rho+1)^{-n} \qquad \text{with } 0 \le i \le n$$

Computing the normalized (with respect to n) limiting probability with $r_1 = 0.8$ and $r_2 = 0.2$ obtained imposing $P_{S_1} = 80\%$ and $r = 1$ in equation (1) we derived the limiting distribution shown in Figure 24 which have the maximum probability in correspondence of 80%. So, the stochastic simulations performed with the model and shown in Figure 7 agree with the hypothesis but also with the theoretical and analytical distribution of the underlying CTMC. Moreover, as the number of states increases, the peak of the limiting probability distribution becomes more smooth confirming that the stochastic simulations are more stable at the equilibrium as the number of simulated processes increases.

Author Index